2

MEDIA AND COMMUNICATIONS - TECHNOLOGIES, POLICIES AND CHALLENGES

BLUETOOTH AND WIRELESS LOCAL AREA NETWORKS

SECURITY GUIDES

MEDIA AND COMMUNICATIONS - TECHNOLOGIES, POLICIES AND CHALLENGES

Additional books in this series can be found on Nova's website under the Series tab.

Additional E-books in this series can be found on Nova's website under the E-book tab.

MEDIA AND COMMUNICATIONS - TECHNOLOGIES, POLICIES AND CHALLENGES

BLUETOOTH AND WIRELESS LOCAL AREA NETWORKS

SECURITY GUIDES

RENE I. JENKINS
EDITOR

NOVINKA

New York

For permission to use material from this book please contact us:
Telephone 631-231-7269; Fax 631-231-8175
Web Site: http://www.novapublishers.com

LIBRARY OF CONGRESS CATALOGING-IN-PUBLICATION DATA

ISBN: 978-1-62417-259-5

Published by Nova Science Publishers, Inc. † New York

CONTENTS

PREFACE

Bluetooth is an open standard for short-range radio frequency communication. Bluetooth technology is used primarily to establish wireless personal area networks (WPANs), and it has been integrated into many types of business and consumer devices. This book provides information on the security capabilities of both Bluetooth technologies and wireless local networks and gives recommendations to organizations on securing them effectively.

Chapter 1 – Bluetooth is an open standard for short-range radio frequency communication. Bluetooth technology is used primarily to establish wireless personal area networks (WPANs), and it has been integrated into many types of business and consumer devices. This publication provides information on the security capabilities of Bluetooth technologies and gives recommendations to organizations employing Bluetooth technologies on securing them effectively. The Bluetooth versions within the scope of this publication are versions 1.1, 1.2, 2.0 + Enhanced Data Rate (EDR), 2.1 + EDR, 3.0 + High Speed (HS), and 4.0, which includes Low Energy (LE) technology.

Chapter 2 – devices within a limited geographic area, such as an office building, that exchange data through radio communications. WLAN technologies are based on the IEEE 802.11 standard and its amendments.[1] The fundamental components of an IEEE 802.11 WLAN (hereafter referred to as a "WLAN" in this publication) are client devices, such as laptops and smartphones, and access points (APs), which logically connect client devices with a distribution system, typically the organization's wired network infrastructure. Some WLANs also use wireless switches, which act as intermediaries between APs and the distribution system.

The security of each WLAN is heavily dependent on how well each WLAN component—including client devices, APs, and wireless switches—is secured throughout the WLAN lifecycle, from initial WLAN design and deployment through ongoing maintenance and monitoring. Unfortunately, WLANs are typically less secure than their wired counterparts for several reasons, including the ease of access to the WLAN and the weak security configurations often used for WLANs (to favor convenience over security). The purpose of this publication is to help organizations improve their WLAN security by providing recommendations for WLAN security configuration and monitoring. This publication supplements other NIST publications by consolidating and strengthening their key recommendations.

In: Bluetooth and Wireless Local Area Networks ISBN: 978-1-62417-259-5
Editor: Rene I. Jenkins

Chapter 1

GUIDE TO BLUETOOTH SECURITY*

John Padgette, Karen Scarfone and Lily Chen

ABSTRACT

Bluetooth is an open standard for short-range radio frequency communication. Bluetooth technology is used primarily to establish wireless personal area networks (WPANs), and it has been integrated into many types of business and consumer devices. This publication provides information on the security capabilities of Bluetooth technologies and gives recommendations to organizations employing Bluetooth technologies on securing them effectively. The Bluetooth versions within the scope of this publication are versions 1.1, 1.2, 2.0 + Enhanced Data Rate (EDR), 2.1 + EDR, 3.0 + High Speed (HS), and 4.0, which includes Low Energy (LE) technology.

Keywords: Bluetooth; information security; network security; wireless networking; wireless personal area networks

EXECUTIVE SUMMARY

Bluetooth is an open standard for short-range radio frequency (RF) communication. Bluetooth technology is used primarily to establish wireless

* This is an edited, reformatted and augmented version of the NIST Publication 800-121 Revision 1, U.S. Department of Commerce, dated June 2012.

personal area networks (WPANs). Bluetooth technology has been integrated into many types of business and consumer devices, including cell phones, laptops, automobiles, medical devices, printers, keyboards, mice, and headsets. This allows users to form ad hoc networks between a wide variety of devices to transfer voice and data. This document provides an overview of Bluetooth technology and discusses related security concerns.

Several Bluetooth versions are currently in use in commercial devices. At the time of writing, Bluetooth 1.2 (adopted November 2003) and 2.0 + Enhanced Data Rate (EDR, adopted November 2004) are the most prevalent. Bluetooth 2.1 + EDR (adopted July 2007), which is quickly becoming the standard, provides significant security improvements for cryptographic key establishment in the form of Secure Simple Pairing (SSP). The most recent versions include Bluetooth 3.0 + High Speed (HS, adopted April 2009), which provides significant data rate improvements, and Bluetooth 4.0 (adopted June 2010), which includes Low Energy (LE) technology that supports smaller, resource-constrained devices and associated applications. This publication addresses the security of all these versions of Bluetooth.

Bluetooth technology and associated devices are susceptible to general wireless networking threats, such as denial of service (DoS) attacks, eavesdropping, man-in-the-middle (MITM) attacks, message modification, and resource misappropriation. They are also threatened by more specific Bluetooth-related attacks that target known vulnerabilities in Bluetooth implementations and specifications. Attacks against improperly secured Bluetooth implementations can provide attackers with unauthorized access to sensitive information and unauthorized use of Bluetooth devices and other systems or networks to which the devices are connected.

To improve the security of Bluetooth implementations, organizations should implement the following recommendations:

Organizations Should Use the Strongest Bluetooth Security Mode That Is Available for Their Bluetooth Devices

The Bluetooth specifications define several security modes, and each version of Bluetooth supports some, but not all, of these modes. The modes differ primarily by the point at which the device initiates security; hence, these modes define how well they protect Bluetooth communications and devices from potential attack. For Bluetooth Basic Rate (BR), EDR, and HS, Security Mode 3 is the strongest mode because it requires establishment of authentication and encryption before the Bluetooth physical link is completely established. However, Security Mode 4 is the default mode for Bluetooth 2.1+EDR and later devices (if both devices support Security Mode 4, then they must use it). Security Modes 2 and 4 can also use authentication and encryption, but do not initiate them until after the Bluetooth physical link has already been fully established and logical channels partially established. Security Mode 1 devices never initiate security and therefore should never be used.

For Bluetooth LE (introduced in Version 4.0), Security Mode 1 Level 3 is considered the strongest mode because it requires authenticated pairing and encryption. Other security modes/levels allow unauthenticated pairing (meaning no man-in-the-middle protection is provided during cryptographic key establishment), and some do not require any security at all.

The available modes vary based on the Bluetooth specification version supported by the device, so organizations should choose the most secure mode available for each case.

Organizations should address Bluetooth technology in their security policies and change default settings of Bluetooth devices to reflect the policies.

A security policy that defines requirements for Bluetooth security is the foundation for all other Bluetooth-related countermeasures. The policy should include a list of approved uses for Bluetooth, a list of the types of information that may be transferred over Bluetooth networks, and requirements for selecting and using Bluetooth personal identification numbers (PINs), where applicable. After establishing a Bluetooth security policy, organizations should ensure that Bluetooth devices' default settings are reviewed and changed as needed so that they comply with the security policy requirements. For example, a typical requirement is to disable unneeded Bluetooth profiles and services to reduce the number of vulnerabilities that attackers could attempt to exploit. When available, a centralized security policy management approach should be used to ensure device configurations are compliant.

Organizations should ensure that their Bluetooth users are made aware of their security-related responsibilities regarding Bluetooth use.

A security awareness program helps educate and train users to follow security practices that protect the assets of an organization and prevent security incidents. For example, users should be provided with a list of precautionary measures they should take to better protect handheld Bluetooth devices from theft. Users should also be made aware of other actions to take regarding Bluetooth device security, such as ensuring that Bluetooth devices are turned off when they are not needed to minimize exposure to malicious activities, and performing Bluetooth device pairing as infrequently as possible and ideally in a physically secure area where attackers cannot observe passkey entry and eavesdrop on Bluetooth pairing-related communications.

1. Introduction

1.1. Purpose and Scope

The purpose of this document is to provide information to organizations on the security capabilities of Bluetooth and provide recommendations to organizations employing Bluetooth technologies on securing them effectively. The Bluetooth versions within the scope of this publication are versions 1.1, 1.2, 2.0 + Enhanced Data Rate (EDR), 2.1 + EDR, 3.0 + High Speed (HS), and 4.0, which includes Low Energy (LE) technology.

1.2. Audience and Assumptions

This document discusses Bluetooth technologies and security capabilities in technical detail. This document assumes that the readers have at least some operating system, wireless networking, and security knowledge. Because of the constantly changing nature of the wireless security industry and the threats and vulnerabilities to the technologies, readers are strongly encouraged to take advantage of other resources (including those listed in this document) for more current and detailed information.

The following list highlights people with differing roles and responsibilities that might use this document:

- Government managers (e.g., chief information officers and senior managers) who oversee the use and security of Bluetooth technologies within their organizations
- Systems engineers and architects who design and implement Bluetooth technologies

- Auditors, security consultants, and others who perform security assessments of wireless environments
- Researchers and analysts who are trying to understand the underlying wireless technologies.

1.3. Document Organization

The remainder of this document is composed of the following sections and appendices:

- Section 2 provides an overview of Bluetooth technology, including its benefits, technical characteristics, and architecture.
- Section 3 discusses the security features defined in the Bluetooth specifications and highlights their limitations.
- Section 4 examines common vulnerabilities and threats involving Bluetooth technologies and makes recommendations for countermeasures to improve Bluetooth security.
- Appendix A provides a glossary of terms.
- Appendix B provides a list of acronyms and abbreviations used in this document.
- Appendix C lists Bluetooth references.
- Appendix D lists Bluetooth online resources.

2. Overview of Bluetooth Technology

Bluetooth is an open standard for short-range radio frequency (RF) communication. Bluetooth technology is used primarily to establish wireless personal area networks (WPANs). Bluetooth technology has been integrated into many types of business and consumer devices, including cell phones, laptops, automobiles, printers, keyboards, mice, and headsets. This allows users to form ad hoc networks between a wide variety of devices to transfer voice and data. Bluetooth is a low-cost, low-power technology that provides a mechanism for creating small wireless networks on an ad hoc basis, known as *piconets*.[1] A piconet is composed of two or more Bluetooth devices in close physical proximity that operate on the same channel using the same frequency hopping sequence. An example of a piconet is a Bluetooth-based connection between a cell phone and a headset.

Bluetooth piconets are often established on a temporary and changing basis, which offers communications flexibility and scalability between mobile devices. Some key benefits of Bluetooth technology are—

- **Cable replacement.** Bluetooth technology replaces a variety of cables, such as those traditionally used for peripheral devices (e.g., mouse and keyboard connections), printers, and wireless headsets and earbuds that interface with desktops, laptops, cell phones, etc.
- **Ease of file sharing**. A Bluetooth-enabled device can form a piconet to support file sharing capabilities with other Bluetooth devices, such as laptops.
- **Wireless synchronization.** Bluetooth can provide automatic synchronization between Bluetooth-enabled devices. For example, Bluetooth allows synchronization of contact information contained in electronic address books and calendars.
- **Internet connectivity.** A Bluetooth device with Internet connectivity can share that access with other Bluetooth devices. For example, a laptop can use a Bluetooth connection to direct a cell phone to establish a dial-up connection so that the laptop can access the Internet through the phone.

Bluetooth technology was originally conceived by Ericsson in 1994. Ericsson, IBM, Intel, Nokia, and Toshiba formed the Bluetooth Special Interest Group (SIG), a not-for-profit trade association developed to drive development of Bluetooth products and serve as the governing body for Bluetooth specifications.[2] Bluetooth is standardized within the IEEE 802.15 Working Group for Wireless Personal Area Networks that formed in early 1999 as IEEE 802.15.1-2002.[3]

This section provides an overview of Bluetooth technology, including frequency and data rates, range, and architecture.

2.1. Bluetooth Technology Characteristics

Bluetooth operates in the unlicensed 2.4000 gigahertz (GHz) to 2.4835 GHz Industrial, Scientific, and Medical (ISM) frequency band. Numerous technologies operate in this band, including the IEEE 802.11b/g wireless local area network (WLAN) standard, making it somewhat crowded from the standpoint of the volume of wireless transmissions. Bluetooth employs frequency hopping spread spectrum (FHSS) technology for transmissions. FHSS reduces interference and transmission errors but provides minimal transmission security. With FHSS technology, communications between Bluetooth BR/EDR devices use 79 different 1 megahertz (MHz) radio channels by hopping (i.e., changing) frequencies about 1,600 times per second for data/voice links and 3,200 times per second during page and inquiry scanning. A channel is used for a very short period (e.g., 625 microseconds for data/voice links), followed by a hop to another channel designated by a pre-determined pseudo-random sequence; this process is repeated continuously in the frequency hopping sequence.

Bluetooth also provides for radio link power control, which allows devices to negotiate and adjust their radio power according to signal strength measurements. Each device in a Bluetooth network can determine its received signal strength indication (RSSI) and request that the other network device adjust its relative radio power level (i.e., incrementally increase or decrease the transmission power). This is performed to conserve power and/or to keep the received signal characteristics within a preferred range.

The combination of a frequency hopping scheme and radio link power control provides Bluetooth with some additional, albeit limited, protection from eavesdropping and malicious access. The frequency-hopping scheme, primarily a technique to avoid interference, makes it slightly more difficult for an adversary to locate and capture Bluetooth transmissions than to capture transmissions from fixed-frequency technologies, like those used in IEEE 802.11b/g. Research published in 2007 has shown that the Bluetooth frequency hopping sequence for an active piconet can be determined using relatively inexpensive hardware and free open source software.[4]

The range of Bluetooth BR/EDR devices is characterized by three classes that define power management. Table 2-1 summarizes the classes, including their power levels in milliwatts (mW) and decibels referenced to one milliwatt (dBm), and their operating ranges in meters (m).[5] Most small, battery-powered devices are Class 2, while Class 1 devices are typically universal serial bus (USB) adapters for desktops and laptops, as well as access points and other mains powered devices.

To allow Bluetooth devices to find and establish communication with each other, discoverable and connectable modes are specified. A device in *discoverable mode* periodically monitors an inquiry scan physical channel (based on a specific set of frequencies) and responds to an inquiry on that channel with its device address, local clock (counter) value, and other characteristics needed to page and subsequently connect to it. A device in *connectable mode* periodically monitors its page scan physical channel and responds to a page on that channel to initiate a network connection. The frequencies associated with the page scan physical channel for a device are based on its Bluetooth device address. Therefore, knowing a device's address and local clock[6] is important for paging and subsequently connecting to the device.

The following sections cover Bluetooth BR/EDR/HS data rates, LE technology, and dual-mode devices.

2.1.1. Basic, Enhanced and High Speed Data Rates

Bluetooth devices can support multiple data rates using native Bluetooth and alternate Medium Access Controls (MAC) and Physical (PHY) Layers. Because Bluetooth specifications are designed to be backward-compatible, a later specification device that supports higher data rates also supports the lower data rates supported by earlier specification devices (e.g., an EDR device also supports rates specified for BR devices). The following sections provide an overview for Bluetooth and alternate MAC/PHYs, as well as associated data rates and modulation schemes.

2.1.1.1. Basic Rate/Enhanced Data Rate

Bluetooth versions 1.1 and 1.2 only support transmission speeds of up to 1 megabit per second (Mbps), which is known as Basic Rate (BR), and can achieve payload throughput of approximately 720 kilobits per second (kbps)

Introduced in Bluetooth version 2.0, Enhanced Data Rate (EDR) specifies data rates up to 3 Mbps and throughput of approximately 2.1 Mbps.

Table 2.1. Bluetooth Device Classes of Power Management

Type	Power	Max Power Level	Designed Operating Range	Sample Devices
Class 1	High	100 mW (20 dBm)	Upto 100 meters (328feet)	USB adapters, access points
Class 2	Medium	2.5 mW (4 dBm)	Upto10 meters (33feet)	Mobiledevices, Bluetooth adapters, smart card readers
Class 3	Low	1mW (0 dBm)	Upto 1 meter (3feet)	Bluetooth adapters

BR uses Gaussian Frequency-Shift Keying (GFSK) modulation to achieve a 1 Mbps data rate. EDR uses π/4 rotated Differential Quaternary Phase Shift Keying (DQPSK) modulation to achieve a 2 Mbps data rate, and 8 phase Differential Phase Shift Keying (8DPSK) to achieve a 3 Mbps data rate.

Note that EDR support is not required for devices compliant with the Bluetooth 2.0 specification or later. Therefore, there are devices on the market that are "Bluetooth 2.0 compliant" versus "Bluetooth 2.0 + EDR compliant." The former are devices that support required version 2.0 features but only provide the BR data rate.

2.1.1.2. High Speed with Alternate MAC/PHY

Introduced in the Bluetooth 3.0 + HS specification, devices can support faster data rates by using Alternate MAC/PHYs (AMP). This is known as Bluetooth High Speed (HS).

In the Bluetooth 3.0 + HS specification, IEEE 802.11-2007 was introduced as the first supported AMP. IEEE 802.11-2007 is a rollup of the amendments IEEE 802.11a through 802.11j. For the 802.11 AMP, IEEE 802.11g PHY support is mandatory, while IEEE 802.11a PHY support is optional. The 802.11 AMP is designed to provide data rates up to 24 Mbps using Orthogonal Frequency-Division Multiplexing (OFDM) modulation.

Note that this AMP is IEEE 802.11 compliant but not Wi-Fi compliant. Therefore, Wi-Fi Alliance specification compliance is not required for Bluetooth 3.0 + HS devices.

2.1.2. Low Energy

Bluetooth LE was introduced in the Bluetooth 4.0 specification. Formerly known as "Wibree" and "Ultra Low Power Bluetooth," LE is primarily designed to bring Bluetooth technology to coin cell battery-powered devices such as medical devices and other sensors. The key technology goals of Bluetooth LE (compared with Bluetooth BR/EDR) include lower power consumption, reduced memory requirements, efficient discovery and connection procedures, short packet lengths, and simple protocols and services.

Table 2.2 provides the key technical differences between BR/EDR and LE.

2.1.3. Dual Mode Devices (Concurrent LE & BR/EDR/HS Support)

A Bluetooth v4.0 device may support both BR/EDR/HS and LE as a "dual mode" Bluetooth device. An example is a cell phone that uses an EDR link to a Bluetooth headset and a concurrent LE link to a sensor that unlocks and starts the user's automobile. Figure 2-1 shows the device architecture for Bluetooth v4.0 devices, and includes BR/EDR, HS and LE technologies. New terms included in the figure related to security are discussed in subsequent sections.

Table 2.2. Key Differences between Bluetooth BR/EDR and LE

Characteristic	Bluetooth BR/EDR	Bluetooth LE
RF Physical Channels	79 channels with 1MHz channel spacing	40 channels with 2MHz channel spacing
Discovery/Connect	Inquiry/Paging	Advertising
Number of Piconet Slaves	7(active)/255(total)	Unlimited
Device Address Privacy	None	Private device addressing available
Max Data Rate	1–3 Mbps	1Mbps via GFSK modulation
Encryption Algorithm	E0/SAFER+	AES-CCM
Typical Range	30 meters	50 meters
Max Output Power	100 mW (20 dBm)	10 mW (10 dBm)

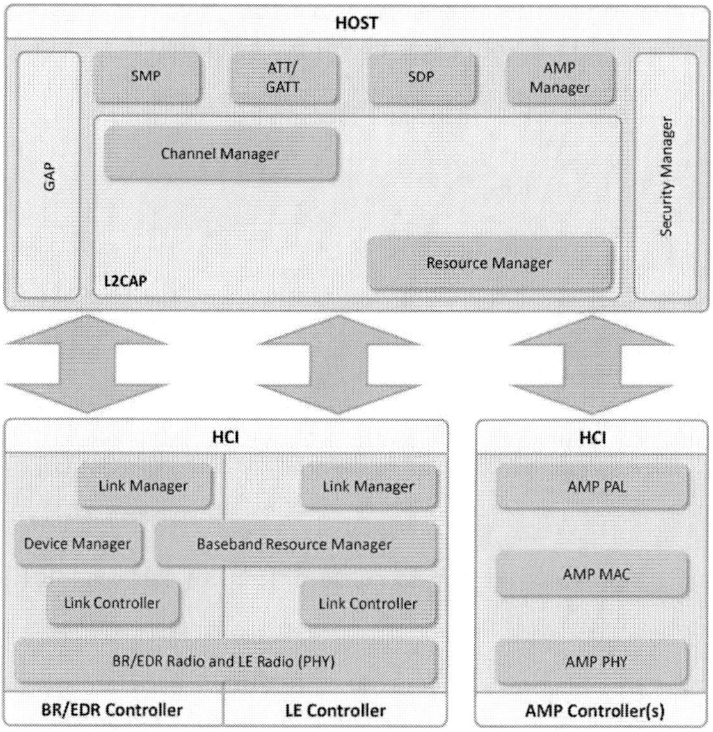

Figure 2.1. Buetooth v4.0 Device Architecture.

2.2. Bluetooth Architecture

Bluetooth permits devices to establish ad hoc networks. Ad hoc networks allow easy connection establishment between devices in the same physical area (e.g., the same room) without the use of any infrastructure devices. A Bluetooth client is simply a device with a Bluetooth radio and software incorporating the Bluetooth protocol stack and interfaces.

The Bluetooth specification provides separation of duties for performing stack functions between a host and a controller. The host is responsible for the higher layer protocols, such as Logical Link Control and Adaptation Protocol (L2CAP) and Service Discovery Protocol (SDP). The host functions are performed by a computing device like a laptop or smartphone. The controller is responsible for the lower layers, including the Radio, Baseband, and Link Control/Management. The controller functions are performed by an integrated or external (e.g., USB) Bluetooth adapter. The host and controller send information to each other using standardized communications over the Host Controller Interface (HCI). This standardized HCI allows hosts and controllers from different product vendors to interoperate. In some cases, the host and controller functions are integrated into a single device; Bluetooth headsets are a prime example.

Figure 2-2 depicts the basic Bluetooth network topology. In a piconet one device serves as the master, with all other devices in the piconet acting as slaves. BR/EDR piconets can scale to include up to 7 active slave devices and up to 255 inactive slave devices. The new Bluetooth LE technology (see Section 2.1.2) allows an unlimited number of slaves.

The master device controls and establishes the network, including defining the network's frequency hopping scheme. Although only one device can serve as the master for each piconet, time division multiplexing (TDM) allows a slave in one piconet to act as the master for another piconet simultaneously, thus creating a chain of networks.[7] This chain, called a *scatternet*, allows networking of several devices over an extended distance in a dynamic topology that can change during any given session. As a device moves toward or away from the master device the topology may change, along with the relationships of the devices in the immediate network. Figure 2-3 depicts a scatternet that involves three piconets.

The Bluetooth core protocols provide no multi-hop network routing capabilities for devices involved in scatternets. For example in Figure 2-3, User C's phone in Piconet 3 cannot communicate with User B's phones in Piconet 2 without establishing an additional piconet between them.

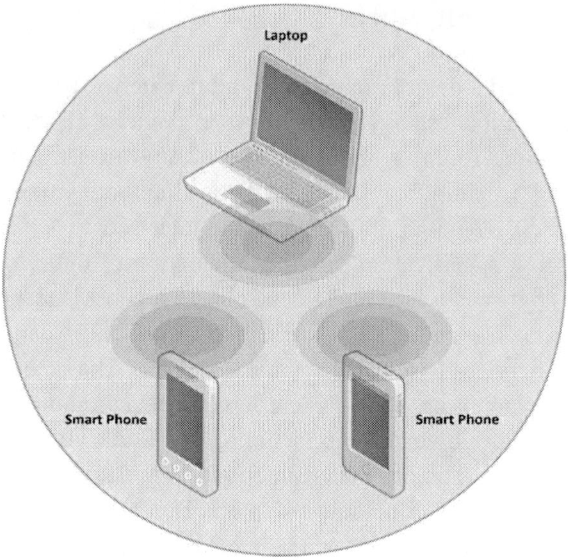

Figure 2.2. Bluetooth Ad Hoc Topology.

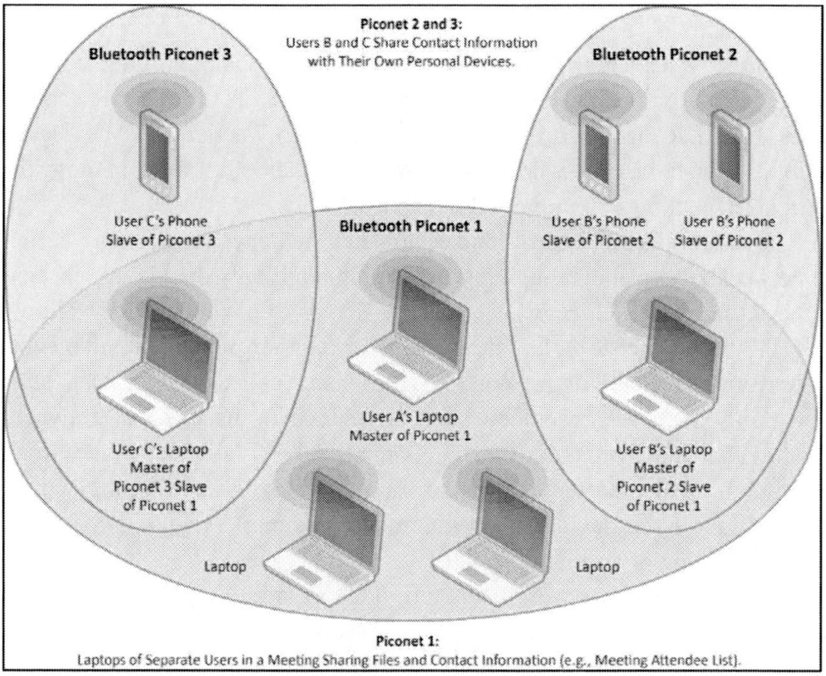

Figure 2.3. Bluetooth Networks (Multiple Scatternets).

Scatternets are only available to BR/EDR devices, because Bluetooth LE technology does not support that feature.

3. Bluetooth Security Features

This section provides an overview of the security mechanisms included in the Bluetooth specifications to illustrate their limitations and provide a foundation for the security recommendations in Section 4. A high-level example of the scope of the security for the Bluetooth radio path is depicted in Figure 3-1. In this example, Bluetooth security is provided between the phone and the laptop, while IEEE 802.11 security protects the WLAN link between the laptop and the IEEE 802.11 AP. Communications on the wired network are not protected by Bluetooth or IEEE 802.11 security capabilities. Therefore, end-to-end security is not possible without using higher-layer security solutions atop the security features included in Bluetooth and IEEE 802.11.

Three basic security services are specified in the Bluetooth standard:

- **Authentication:** verifying the identity of communicating devices based on their Bluetooth device address. Bluetooth does not provide native user authentication.
- **Confidentiality:** preventing information compromise caused by eavesdropping by ensuring that only authorized devices can access and view transmitted data.
- **Authorization:** allowing the control of resources by ensuring that a device is authorized to use a service before permitting it to do so.

The three security services offered by Bluetooth and details about the modes of security are described below. Bluetooth does not address other security services such as audit, integrity, and non-repudiation; if such services are needed, they should be provided through additional means.

3.1. Security Features of Bluetooth BR/EDR/HS

Cumulatively, the family of Bluetooth BR/EDR/HS specifications defines four security modes. Each Bluetooth device must operate in one of these modes, called Security Modes 1 through 4. These modes dictate when a Bluetooth device initiates security, not whether it supports security features.

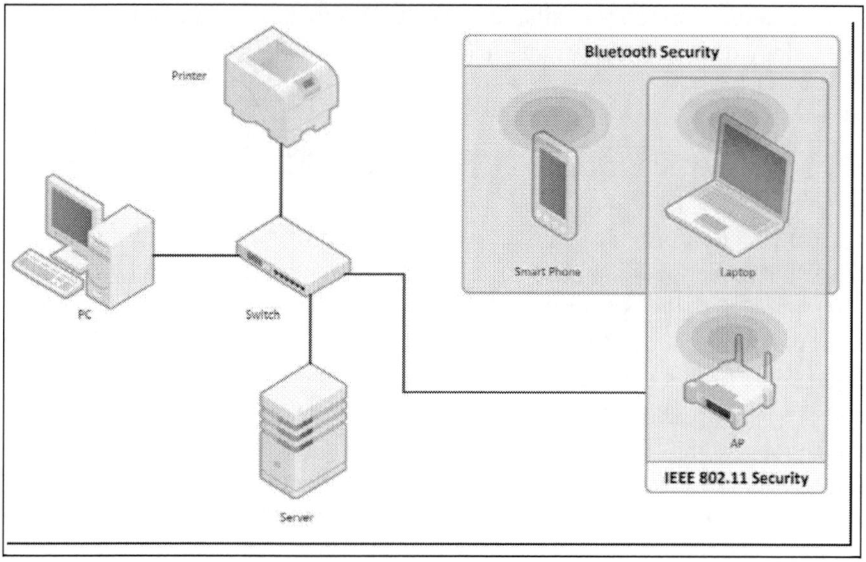

Figure 3.1. Bluetooth Air-Interface Security.

Security Mode 1 devices are considered non-secure. Security functionality (authentication and encryption) is never initiated, leaving the device and connections susceptible to attackers.

In effect, Bluetooth devices in this mode are "indiscriminate" and do not employ any mechanisms to prevent other Bluetooth-enabled devices from establishing connections. However, if a remote device initiates security—such as a pairing, authentication, or encryption request—a Security Mode 1 device will participate. Per their respective Bluetooth specification versions, all v2.0 and earlier devices can support Security Mode 1, and v2.1 and later devices can use Security Mode 1 for backward compatibility with older devices. However, NIST recommends never using Security Mode 1.

In Security Mode 2, a service level-enforced security mode, security procedures may be initiated after link establishment but before logical channel establishment. For this security mode, a local security manager (as specified in the Bluetooth architecture) controls access to specific services. The centralized security manager maintains policies for access control and interfaces with other protocols and device users.

Varying security policies and trust levels to restrict access can be defined for applications with different security requirements operating in parallel. It is possible to grant access to some services without providing access to other services. In this mode, the notion of authorization—the process of deciding whether a specific device is allowed to have access to a specific service—is introduced. Typically Bluetooth service discovery can be performed prior to any security challenges (i.e., authentication, encryption, and/or authorization). However, all other Bluetooth services should require all of those security mechanisms.

It is important to note that the authentication and encryption mechanisms used for Security Mode 2 are implemented in the controller, as with Security Mode 3 described below. All v2.0 and earlier devices can support Security Mode 2, but v2.1 and later devices can only support it for backward compatibility with v2.0 or earlier devices.

Security Mode 3 is the link level-enforced security mode, in which a Bluetooth device initiates security procedures before the physical link is fully established.

Bluetooth devices operating in Security Mode 3 mandate authentication and encryption for all connections to and from the device. Therefore, even service discovery cannot be performed until after authentication, encryption, and authorization have been performed.

Once a device has been authenticated, service-level authorization is not typically performed by a Security Mode 3 device. However, NIST recommends that service-level authorization should be performed to prevent "authentication abuse"—that is, an authenticated remote device using a Bluetooth service without the local device owner's knowledge.

All v2.0 and earlier devices can support Security Mode 3, but v2.1 and later devices can only support it for backward compatibility purposes.

Similar to Security Mode 2, Security Mode 4 (introduced in Bluetooth v2.1 + EDR) is a service-level-enforced security mode in which security procedures are initiated after physical and logical link setup.

Security Mode 4 uses Secure Simple Pairing (SSP), in which Elliptic Curve Diffie-Hellman (ECDH) key agreement replaces legacy key agreement for link key generation (see Section 3.1.1). However, the device authentication and encryption algorithms are identical to the algorithms in Bluetooth v2.0 + EDR and earlier versions.

Security requirements for services protected by Security Mode 4 must be classified as one of the following:

- Authenticated link key required
- Unauthenticated link key required
- No security required.

Whether or not a link key is authenticated depends on the SSP association model used (see Section 3.1.1.2). Security Mode 4 requires encryption for all services (except Service Discovery) and is mandatory for communication between v2.1 and later BR/EDR devices. However, for backward compatibility, a Security Mode 4 device can fall back to any of the other three Security Modes when communicating with Bluetooth v2.0 and earlier devices that do not support Security Mode 4. In this case, NIST recommends using Security Mode 3.

The remainder of this section discusses specific Bluetooth security components in more detail—pairing and link key generation, authentication, confidentiality, and other Bluetooth security features.

3.1.1. Pairing and Link Key Generation

Essential to the authentication and encryption mechanisms provided by Bluetooth is the generation of a secret symmetric key, called the ^{3}Oink key.′ As mentioned in Section 3.1, Bluetooth BR/EDR performs pairing (i.e., link key generation) in one of two ways. Security Modes 2 and 3 initiate link key establishment via a method called Personal Identification Number (PIN) Pairing (i.e., Legacy or Classic Pairing), while Security Mode 4 uses SSP. Both methods are described below.

3.1.1.1. PIN/Legacy Pairing

For PIN/legacy pairing, two Bluetooth devices simultaneously derive link keys when the user(s) enter an identical secret PIN into one or both devices, depending on the configuration and device type. The PIN entry and key derivation are depicted conceptually in Figure 3-2. Note that if the PIN is less than 16 bytes, the initiating device's address (BD_ADDR) supplements the PIN value to generate the initialization key. The E_x boxes represent encryption algorithms that are used during the Bluetooth link key derivation processes. More details on the Bluetooth authentication and encryption procedures are outlined in Sections 3.1.2 and 3.1.3, respectively.

After link key generation is complete, the devices complete pairing by mutually authenticating each other to verify they have the same link key. The PIN code used in Bluetooth pairing can vary between 1 and 16 bytes of binary or, more commonly, alphanumeric characters.

The typical four-digit PIN may be sufficient for low-risk situations; a longer PIN (e.g., 8-character alphanumeric) should be used for devices that require a higher level of security.[8]

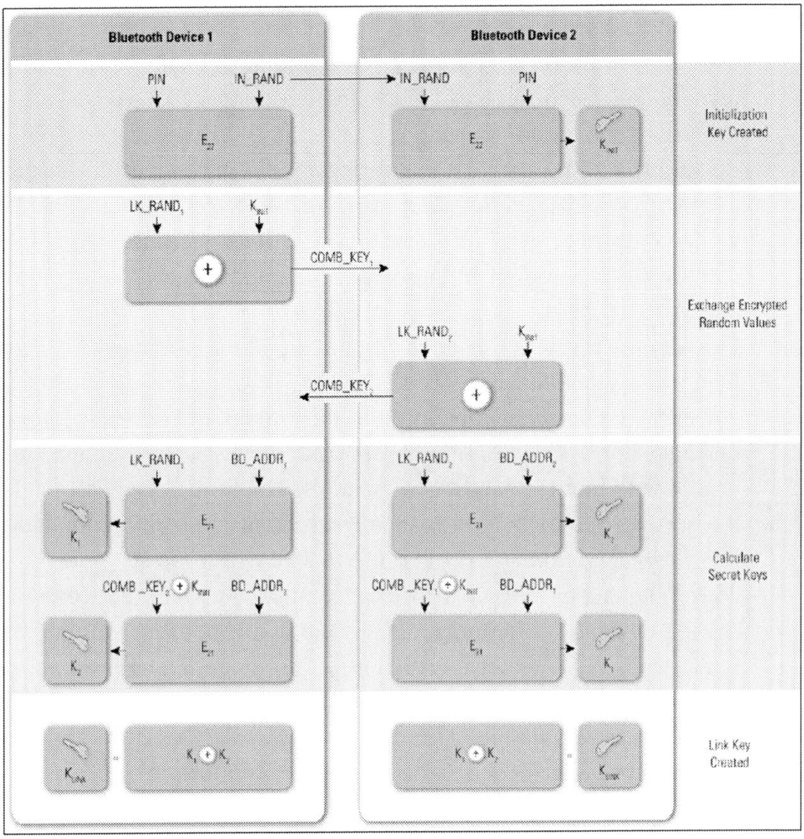

Figure 3.2. Link Key Generation from PIN.

3.1.1.2. Secure Simple Pairing

SSP was introduced in Bluetooth v2.1 + EDR for use with Security Mode 4. SSP simplifies the pairing process by providing a number of association models that are flexible in terms of device input/output capability. SSP also improves security through the addition of ECDH public key cryptography for protection against passive eavesdropping and man-in-the-middle attacks (MITM) during pairing.

The four association models offered in SSP are as follows:[9]

- **Numeric Comparison** was designed for the situation where both Bluetooth devices are capable of displaying a six-digit number and allowing a user to enter a "yes" or "no" response. During pairing, a user is shown a six-digit number on each display and provides a "yes" response on each device if the numbers match. Otherwise, the user responds "no" and pairing fails. A key difference between this operation and the use of PINs in legacy pairing is that the displayed number is not used as input for link key generation. Therefore, an eavesdropper who is able to view (or otherwise capture) the displayed value could not use it to determine the resulting link or encryption key.

- **Passkey Entry** was designed for the situation where one Bluetooth device has input capability (e.g., keyboard), while the other device has a display but no input capability. In this model, the device with only a display shows a six-digit number that the user then enters on the device with input capability. As with the Numeric Comparison model, the six-digit number used in this transaction is not incorporated into link key generation and is of no use to an eavesdropper.
- **Just Works** was designed for the situation where at least one of the pairing devices has neither a display nor a keyboard for entering digits (e.g., headset). It performs Authentication Stage 1 (see Figure 3-3) in the same manner as the Numeric Comparison model, except that a display is not available. The user is required to accept a connection without verifying the calculated value on both devices, so Just Works provides no MITM protection.
- **Out of Band (OOB)** was designed for devices that support a common additional wireless or wired technology (e.g., Near Field Communication or NFC) for the purposes of device discovery and cryptographic value exchange. In the case of NFC, the OOB model allows devices to pair by simply "tapping" one device against the other, followed by the user accepting the pairing via a single button push. It is important to note that to keep the pairing process as secure as possible, the OOB technology should be designed and configured to mitigate eavesdropping and MITM attacks.

Security Mode 4 requires Bluetooth services to mandate an authenticated link key, an unauthenticated link key, or no security at all. Of the association models described above, all but the Just Works model provide authenticated link keys. Figure 3.3 shows how the link key is established for SSP. Note how this technique uses ECDH public/private key pairs rather than generating a symmetric key via a PIN.

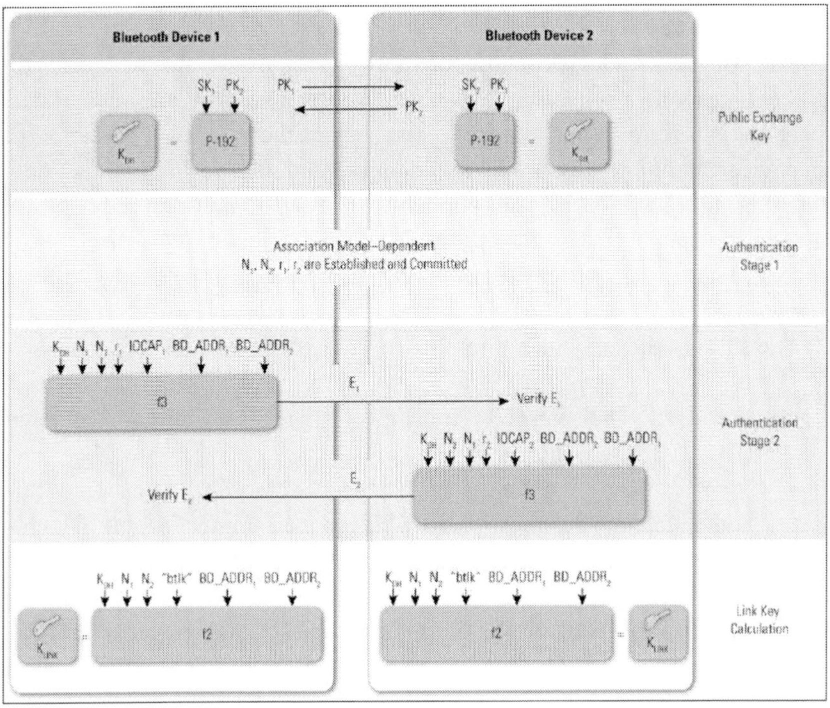

Figure 3.3. Link Key Establishment for Secure Simple Pairing.

3.1.1.3. AMP Link Key Derivation from Bluetooth Link Key

For AMP link security (e.g., IEEE 802.11, as introduced in Bluetooth v3.0), an AMP link key is derived from the Bluetooth link key. A Generic AMP Link Key (GAMP_LK) is generated by the AMP Manager in the host stack whenever a Bluetooth link key is created or changed.

As shown in Figure 3-4, the GAMP_LK is generated using the Bluetooth link key (concatenated with itself) and an extended ASCII key identifier (keyID) of "gamp" as inputs to a HMAC-SHA-256 function. Subsequently, a Dedicated AMP Link Key (for a specific AMP and Trusted Device combination) is derived from the Generic AMP Link Key and keyID. For the 802.11 AMP Link Key, the keyID is "802b".

For IEEE 802.11 AMPs, the Dedicated AMP Link Key is used as the 802.11 Pairwise Master Key. See NIST Special Publication 800-97, *Establishing Wireless Robust Security Networks: A Guide to IEEE 802.11i*[10], for more information about IEEE 802.11 security.

3.1.2. Authentication

The Bluetooth device authentication procedure is in the form of a challenge–response scheme. Each device interacting in an authentication procedure is referred to as either the claimant or the verifier. The *claimant* is the device attempting to prove its identity, and the *verifier* is the device validating the identity of the claimant. The challenge–response protocol validates devices by verifying the knowledge of a secret key—the Bluetooth link key. Figure 3-5 conceptually depicts the challenge–response verification scheme.

The steps in the authentication process are as follows:

- **Step 1.** The verifier transmits a 128-bit random challenge (AU_RAND) to the claimant.
- **Step 2.** The claimant uses the E_1 algorithm[11] to compute an authentication response using his or her unique 48-bit Bluetooth device address (BD_ADDR), the link key, and AU_RAND as inputs. The verifier performs the same computation. Only the 32 most significant bits of the E_1 output are used for authentication purposes. The remaining 96 bits of the 128-bit output are known as the Authenticated Ciphering Offset (ACO) value, which will be used later as input to create the Bluetooth encryption key.
- **Step 3.** The claimant returns the most significant 32 bits of the E_1 output as the computed response, the Signed Response (SRES), to the verifier.
- **Step 4.** The verifier compares the SRES from the claimant with the value that it computed.
- **Step 5.** If the two 32-bit values are equal, the authentication is considered successful. If the two 32-bit values are not equal, the authentication fails.

Performing these steps once accomplishes one-way authentication. The Bluetooth standard allows both one-way and mutual authentication to be performed. For mutual authentication, the above process is repeated with the verifier and claimant switching roles. If authentication fails, a Bluetooth device waits an interval of time before making a new attempt. This time interval increases exponentially to prevent an adversary from attempting to gain access by defeating the authentication scheme through trial-and-error with different link keys. It is important to note that this technique does not provide security against offline attacks to determine the link key using eavesdropped pairing frames and exhaustively guessing PINs.

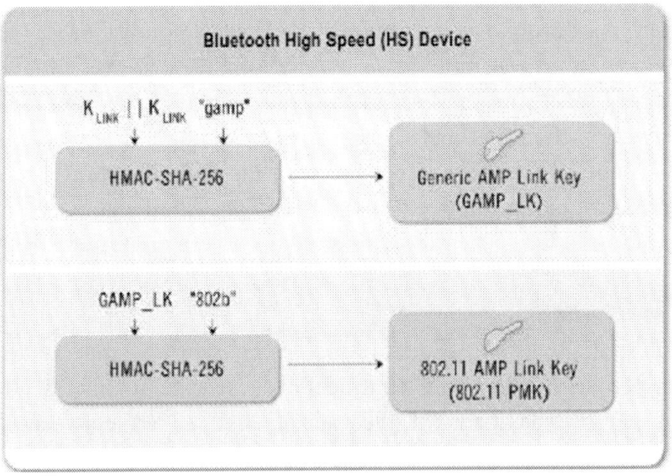

Figure 3.4. AMP Link Key Derivation.

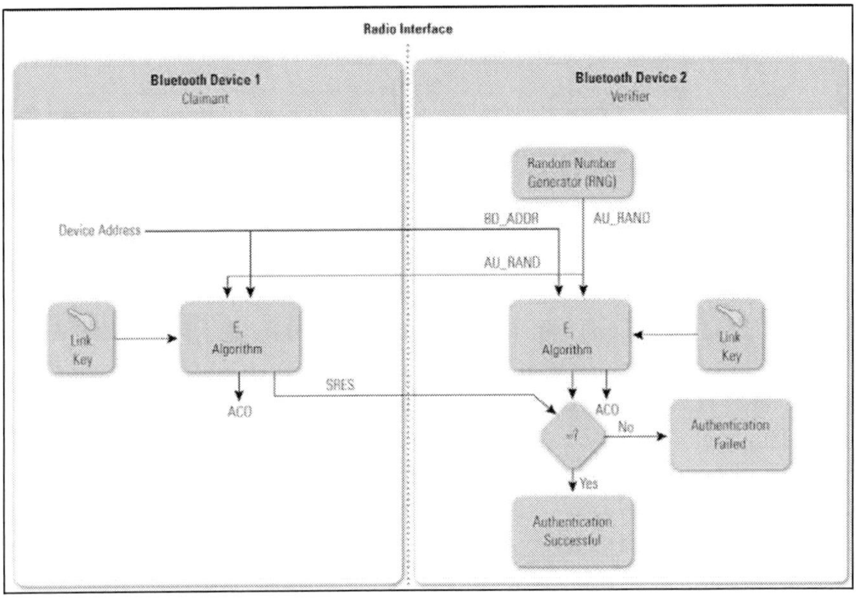

Figure 3.5. Bluetooth Authentication.

Note that the security associated with authentication is solely based on the secrecy of the link key. While the Bluetooth device addresses and random challenge value are considered public parameters, the link key is not. The link key is derived during pairing and should never be disclosed outside the Bluetooth device or transmitted over wireless links. However, the link key is passed in the clear from the host to the controller (e.g., PC to USB adapter) and the reverse when the host is used for key storage. The challenge value, which is a public parameter associated with the authentication process, must be random and unique for every transaction. The challenge value is derived from a pseudo-random generator within the Bluetooth controller.

3.1.3. Confidentiality

In addition to the Security Modes for pairing and authentication, Bluetooth provides a separate confidentiality service to thwart attempts to eavesdrop on the payloads of the packets exchanged between Bluetooth devices. Bluetooth has three Encryption Modes, but only two of them actually provide confidentiality. The modes are as follows:

- **Encryption Mode 1**—No encryption is performed on any traffic.
- **Encryption Mode 2**—Individually addressed traffic is encrypted using encryption keys based on individual link keys; broadcast traffic is not encrypted.
- **Encryption Mode 3**—All traffic is encrypted using an encryption key based on the master link key.

Encryption Modes 2 and 3 use the same encryption mechanism.

Security Mode 4 introduced in Bluetooth 2.1 + EDR requires that encryption be used for all data traffic, except for service discovery.

As shown in Figure 3-6, the encryption key provided to the encryption algorithm is produced using an internal key generator (KG). The KG produces stream cipher keys based on the 128-bit link key, which is a secret that is held in the Bluetooth devices; a 128-bit random number (EN_RAND); and the 96-bit ACO value. The ACO is produced during the authentication procedure, as shown in Figure 3-4.

The Bluetooth encryption procedure is based on a stream cipher, E_0. A key stream output is *exclusive-OR-ed* with the payload bits and sent to the receiving device. This key stream is produced using a cryptographic algorithm based on linear feedback shift registers (LFSRs).[12] The encryption function takes the following as inputs: the master device address (BD_ADDR), the 128-

bit random number (EN_RAND), a slot number based on the piconet clock, and an encryption key, which when combined initialize the LFSRs before the transmission of each packet, if encryption is enabled. The slot number used in the stream cipher changes with each packet; the ciphering engine is also reinitialized with each packet while the other variables remain static.

The encryption key (K_C) is derived from the current link key and may vary in length in single byte increments from 1 byte to 16 bytes in length, as set during a negotiation process that occurs between the master and slave devices. During this negotiation, a master device makes a key size suggestion for the slave. The initial key size suggested by the master is programmed into the controller by the manufacturer and is not always 16 bytes. In product implementations, a "minimum acceptable" key size parameter can be set to prevent a malicious user from driving the key size down to the minimum of 1 byte, which would make the link less secure.

It is important to note that E_0 is not a Federal Information Processing Standards (FIPS) approved algorithm and has come under scrutiny in terms of algorithmic strength.[13] A published theoretical known-plaintext attack can recover the encryption key in 2^{38} computations, compared with a brute force attack, which would require testing 2^{128} possible keys. If communications require FIPS-approved cryptographic protection (e.g., to protect sensitive information transmitted by Federal agencies), this protection can be achieved by layering application-level FIPS-approved encryption over the native Bluetooth encryption.

3.1.4. Trust Levels, Service Security Levels, and Authorization

In addition to the four security modes, Bluetooth allows different levels of trust and service security.

The two Bluetooth levels of trust are trusted and untrusted. A *trusted device* has a fixed relationship with another device and has full access to all services. An *untrusted device* does not have an established relationship with another Bluetooth device, which results in the untrusted device receiving restricted access to services.

Available service security levels depend on the security mode being used. For Security Modes 1 and 3, no service security levels are specified. For Security Mode 2, the following security requirements can be enforced:

- Authentication required
- Encryption required
- Authorization required

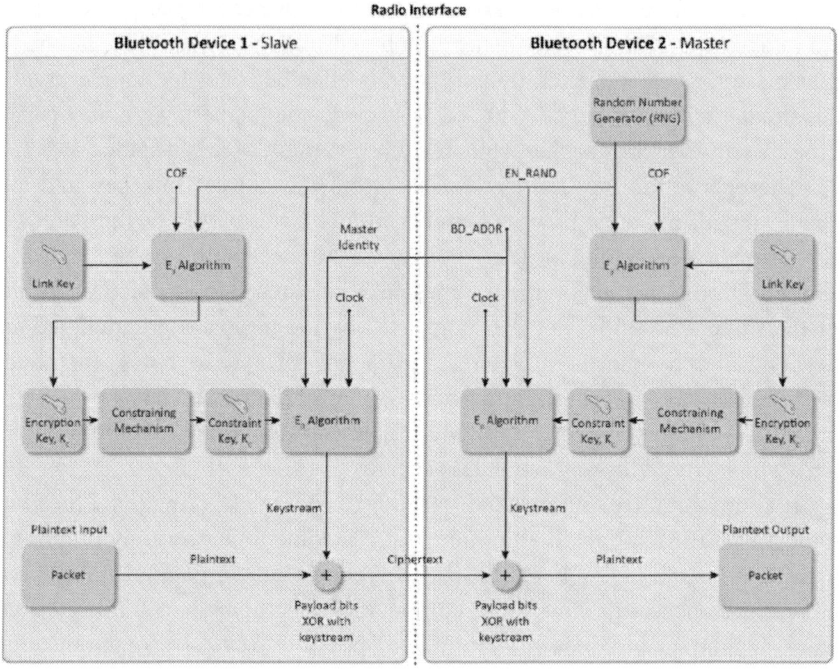

Figure 3.6. Bluetooth Encryption Procedure.

Thus, the available service security levels include any combination of the above, including the lack of security (typically only used for service discovery). Note that BR/EDR encryption cannot be performed without authentication, because the encryption key is derived from an artifact of the authentication process (see Section 3.1.3).

For Security Mode 4, the Bluetooth specification specifies four levels of security for Bluetooth services for use during SSP. The service security levels are as follows:

- **Service Level 3**—Requires MITM protection and encryption; user interaction is acceptable.
- **Service Level 2**—Requires encryption only; MITM protection is not necessary.
- **Service Level 1**—MITM protection and encryption not required. Minimal user interaction.
- **Service Level 0**—No MITM protection, encryption, or user interaction required.

The Bluetooth architecture allows for defining security policies that can set trust relationships in such a way that even trusted devices could gain access only to specific services. Although Bluetooth core protocols can only authenticate devices and not users, user-based authentication is still possible. The Bluetooth security architecture (through the security manager) allows applications to enforce more granular security policies. The link layer at which Bluetooth-specific security controls operate is transparent to the security controls imposed by the application layers. Thus, user-based authentication and fine-grained access control within the Bluetooth security framework are possible through the application layers, although doing so is beyond the scope of the Bluetooth specification.

3.2. Security Features of Bluetooth LE

Because of the intent for Bluetooth LE to support computationally and storage-constrained devices, LE security is different from Bluetooth BR/EDR/HS. One difference is that LE pairing results in the generation of a Long-Term Key (LTK) rather than a Link Key. While fundamentally performing the same secret key function as the Link Key, the LTK is established in a different manner. The LTK is generated using a key transport protocol rather than key agreement as with BR/EDR. That is, one device determines the LTK and securely sends it over to the other device during pairing—instead of both devices generating the same key individually.

LE introduces the use of Advanced Encryption Standard–Counter with CBC-MAC (AES-CCM) encryption for the first time in a Bluetooth specification. In addition to providing strong, standards-based encryption, the inclusion of AES-CCM paves the way for native FIPS-140 validation of Bluetooth LE devices in the future.

LE also introduces features such as private device addresses and data signing. New cryptographic keys called the Identity Resolving Key (IRK) and Connection Signature Resolving Key (CSRK) support these features, respectively.

With LE's privacy feature enabled, the IRK is used to resolve private to public device address mapping. This allows a trusted device to determine another device's public device address from a periodically-changing private device address. Previously, a device would be assigned a static "public" address that would be made available during discovery. If that device remained discoverable, its location could easily be tracked by an adversary.

The use of a periodically-changing private address (an IRK-encrypted form of the public address) mitigates this threat. Since a discoverable LE device transmits ("advertises") identity information, this privacy feature is especially useful.

The CSRK is used to verify cryptographically signed data frames from a particular device. This allows a Bluetooth connection to use data signing (providing integrity and authentication) to protect the connection instead of data encryption (which, in the case of AES-CCM, provides confidentiality, integrity, and authentication).

All of these cryptographic keys (i.e., LTK, IRK, CSRK) are generated and securely distributed during LE pairing. See Section 3.2.2 for details.

3.2.1. LE Security Modes and Levels

LE security modes are similar to BR/EDR service-level security modes (i.e., Security Modes 2 and 4) in that each service can have its own security requirements. However, Bluetooth LE also specifies that each service request can have its own security requirements as well. A device enforces the service-related security requirements by following the appropriate security mode and level.

- LE Security Mode 1 has multiple levels associated with encryption. Level 1 specifies no security, meaning no authentication and no encryption will be initiated. Level 2 requires unauthenticated pairing with encryption. Level 3 requires authenticated pairing with encryption.
- LE Security Mode 2 has multiple levels associated with data signing. Data signing provides strong data integrity but not confidentiality. Level 1 requires unauthenticated pairing with data signing. Level 2 requires authenticated pairing with data signing.

If a particular service request and the associated service have different security modes and/or levels, the stronger security requirements prevail. For example, if either requires Security Mode 1 Level 3, then the requirements for Security Mode 1 Level 3 are enforced.

Because Security Mode 1 Level 3 requires authenticated pairing and encryption, NIST considers this the most secure of these modes/levels and strongly recommends its use for all LE connections. Security Mode 1 Level 1 is the least secure and should never be used. Also, because Security Mode 2 does not provide encryption, Security Mode 1 Level 3 is strongly preferred over Security Mode 2.

3.2.2. LE Pairing Methods

Although LE uses similar pairing method names to BR/EDR SSP, LE pairing does not use ECDH-based cryptography and provides no eavesdropping protection. Therefore, if an attacker can capture the LE pairing frames, he/she may be able to determine the resulting LTK.

Because key transport is used rather than key agreement for LE pairing, a key distribution step is required during LE pairing.

As shown in Figure 3-7, LE pairing begins with the two devices agreeing on a Temporary Key (TK), whose value depends on the pairing method being used. The devices then exchange random values and generate a Short Term Key (STK) based on these values and the TK. The link is then encrypted using the STK, which allows secure key distribution of the LTK, IRK, and CSRK.

The following subsections describe the LE pairing association models. As with BR/EDR SSP, the association model that is used for a particular connection is based on the input/output capabilities of both devices.

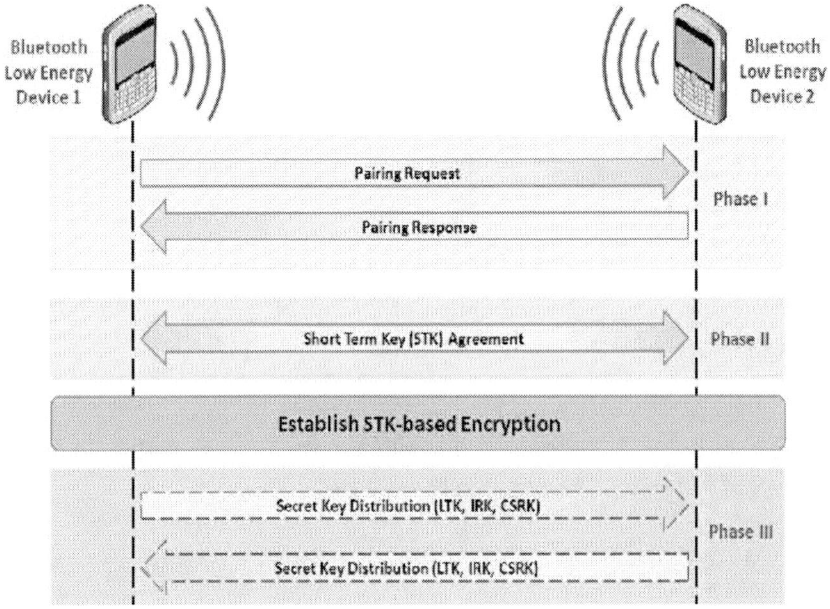

Figure 3.7. Bluetooth Low Energy Pairing.

It is important to note that the LE pairing association model names are similar to those from BR/EDR SSP association models, but the security the models provide are very different.

3.2.2.1. Out of Band

If both devices support a common OOB technology, such as NFC or tethering, they will use the OOB method to pair. In this model, the TK is passed over the OOB technology from one device to the other.

The TK should be unique, random, and equivalent to six decimal digits (i.e., in the hexadecimal range 0x0–0xF423F) at a minimum. NIST strongly recommends use of a full 128-bit random binary (non-alphanumeric) value when practical.

Because OOB pairing results in an authenticated LTK, it should provide at least one-in-a-million protection against MITM attacks—based on the premise that an attacker would have to successfully guess the six-digit TK value. However, the actual protection provided by OOB pairing depends on the MITM protection provided by the OOB technology itself because a successful OOB eavesdropper would know the TK value instead of having to guess it.

3.2.2.2. Passkey Entry

If the devices do not support a common OOB technology, the pairing method to be used is determined based on the input/output capabilities of both devices. If, at a minimum, one device supports keyboard input and the other a display output (or keyboard input as well), then the Passkey Entry pairing method is used to pair.

In this model, the TK is generated from the passkey generated and/or entered in each device. The specification requires the passkey size to be 6 numeric digits; therefore, a maximum of 20 bits of entropy can be provided. Passkey Entry pairing also results in an authenticated LTK.

Because a six-digit passkey is used, an attacker would have a one-in-a-million chance of guessing the correct passkey to perform a MITM attack. NIST recommends using a unique, random passkey for each pairing to provide this level of protection across multiple pairings.

3.2.2.3. Just Works

If neither OOB nor Passkey Entry association models are possible because of device input/output limitations, then the Just Works pairing method is used. As with SSP in BR/EDR/HS, the Just Works pairing method for LE is the weakest of the pairing options from a security perspective.

In this model, the TK is set to all zeros (0x00). Therefore, an eavesdropper or MITM attacker does not need to guess the TK to generate the STK. The Just Works pairing method results in an unauthenticated LTK because no MITM protection is provided during pairing.

3.2.3. LE Key Generation and Distribution

Once the link is encrypted using the STK, the two devices distribute secret keys such as LTK, IRK, and CSRK. Two options are specified for key generation prior to distribution. A device may simply generate random 128-bit values and store them in a local database (called "Database Lookup" in the specification).

The other option is to use a single 128-bit static but random value called Encryption Root (ER) along with a 16-bit Diversifier (DIV) unique to each trusted device to generate the keys.

This option is called "Key Hierarchy" in the specification. For example, the keys can be derived from ER, DIV, and the Identity Root (IR) using the following formulas:

$$LTK = d1(ER, DIV, 0)$$

$$CSRK = d1(ER, DIV, 1)$$

$$IRK = d1(IR, 1, 0)$$

The d1 function is a called a Diversifying Function and is based on AES-128 encryption. However, the specification allows the user of other key derivation functions (e.g., those discussed in NIST SP 800-108, *Recommendation for Key Derivation Using Pseudorandom Functions*[14]).

Using this Key Hierarchy method, the device does not need to store multiple 128-bit keys for each trusted device; rather, it only needs to store its ER and the unique DIVs for each device.

During reconnection, the remote device sends its EDIV, which is a masked version of DIV.[15] The local device can then regenerate the LTK and/or CSRK from its ER and the passed EDIV. If data encryption or signing is set up successfully, it is verified that the remote device had the correct LTK or CSRK.

If unsuccessful, the link is dropped. Note in the above example that the IRK is static and device-specific and therefore could be generated prior to pairing (e.g., during manufacturing).

3.2.4. Confidentiality, Authentication, and Integrity

AES-CCM is used in Bluetooth LE to provide confidentiality as well as per-packet authentication and integrity. There is no separate authentication challenge/response step as with BR/EDR/HS to verify that they both have the same LTK or CSRK.

Because the LTK is used as input for the encryption key, successful encryption setup provides implicit authentication. Similarly, data signing provides implicit authentication that the remote device holds the correct CSRK—although confidentiality is not provided.

4. Bluetooth Vulnerabilities, Threats, and Countermeasures

This section describes vulnerabilities in Bluetooth technologies and threats against those vulnerabilities. Based on these identified common vulnerabilities and threats, as well as the Bluetooth security features described in Section 3, this section also recommends possible countermeasures that can be used to improve Bluetooth security.

Organizations that are planning countermeasures for Bluetooth technologies that use the Bluetooth v4.0 specification should carefully consider its security implications. The specification was released in mid-2010, and at the time of this writing, few products that support the specification are available for evaluation.

As compliant products become available, additional vulnerabilities will likely be discovered, and additional recommendations will be needed for effectively securing Bluetooth LE devices.

Organizations planning to deploy Bluetooth LE devices should carefully monitor developments involving new vulnerabilities, threats, and additional security control recommendations.

4.1. Bluetooth Vulnerabilities

Table 4-1 provides an overview of a number of known security vulnerabilities associated with Bluetooth. The Bluetooth security checklist in Section 4.4 addresses these vulnerabilities.

Table 4.1. Key Problems with Native Bluetooth Security

	Security Issue or Vulnerability	Remarks
Versions Before Bluetoothv1.2		
1	Link keys based on unit keys are static and reused for every pairing.	A device that uses unit keys will use the same link key for every device with which it pairs. This is a serious cryptographic key management vulnerability.
2	Use of link keys based on unit keys can lead to eavesdropping and spoofing.	Once a device's unit key is divulged (i.e., upon its first pairing), any other device that has the key can spoof that device or any other device with which it has paired. Further, it can eavesdrop on that device's connections whether they are encrypted or not.
Versions Before Bluetooth v2.1		
3	Security Mode 1 devices never initiate security mechanisms.	Devices that use Security Mode 1 are inherently insecure. For v2.0 and earlier devices, Security Mode 3 (link level security) is highly recommended.
4	PINs can be too short.	Weak PINs, which are used to protect the generation of link keys during pairing, can be easily guessed. People have a tendency to select short PINs.
5	PIN management and randomness is lacking.	Establishing use of adequate PINs in an enterprise setting with many users may be difficult. Scalability problems frequently yield security problems. The best alternative is for one of the devices being paired to generate the PIN using its random number generator.
6	The encryption keystream repeats after 23.3 hours of use.	As shown in Figure 3-6, the encryption keystream is dependent on the link key, EN_RAND, Master BD_ADDR, and Clock. Only the Master's clock will change during a particular encrypted connection. If a connection lasts more than 23.3 hours, the clock value will begin to repeat, hence generating an identical keystream to that used earlier in the connection. Repeating a keystream is a serious cryptographic vulnerability that would allow an attacker to determine the original plaintext.
Bluetooth v2.1 and v3.0		
7	Just Works association model does not provide	For highest security, devices should require MITM protection during SSP and refuse to accept
	MITM protection during pairing, which results in an unauthen-ticated link key.	unauthenticated link keys generated using Just Works pairing.
8	SSP ECDH key pairs may be static or otherwise weakly generated.	Weak ECDH key pairs minimize SSP eavesdropping protection, which may allow attackers to determine secret link keys. All devices should have unique, strongly-generated ECDH key pairs that change regularly.
9	Static SSP passkeys facilitate MITM attacks.	Passkeys provide MITM protection during SSP. Devices should use random, unique passkeys for each pairing attempt.

Table 4.1. (Continued)

	Security Issue or Vulnerability	Remarks
10	Security Mode 4 devices (i.e.,v2.1 or later) are allowed to fall back to any other security mode when connecting with devices that do not support Security Mode 4 (i.e.,v2.0 and earlier).	The worst-case scenario would be a device falling back to Security Mode 1, which provides no security. NIST strongly recommends that a Security Mode 4 device fall back to Security Mode 3 in this scenario.
Versions Before Bluetooth v4.0		
11	Attempts for authentication are repeatable.	A mechanism needs to be included in Bluetooth devices to prevent unlimited authentication requests. The Blue-tooth specification requires an exponentially increasing waiting interval between successive authentication attempts. However, it does not require such a waiting interval for authentication challenge requests, so an attacker could collect large numbers of challenge responses (which are encrypted with the secret link key) that could leak information about the secret link key.
12	The master key used for broadcast encryption is shared among all piconet devices.	Secret keys shared amongst more than two parties facilitate impersonation attacks.
13	The E0 stream cipher algorithm used for Blue-tooth BR/EDR encrypt-tion is relatively weak.	FIPS-approved encryption can be achieved by layering application-level FIPS-approved encryption over the Bluetooth BR/EDR encryption. Note that Bluetooth LE uses AES-CCM.
14	Privacy may be compro-mised if the Bluetooth device address (BD_ADDR) is captured and associated with a particular user.	Once the BD_ADDR is associated with a particular user, that user's activities and location could be tracked.
15	Device authentication is simple shared-key challenge/response.	One-way-only challenge/response authentication is subject to MITM attacks. Bluetooth provides for mutual authentication, which should be used to provide verification that devices are legitimate.
Bluetooth v4.0		
16	LE pairing provides no eavesdropping protection. Further, the Just Works pairing method provides no MITM protection.	If successful, eavesdroppers can capture secret keys (i.e., LTK, CSRK, IRK) distributed during LE pairing. Further, MITM attackers can capture and manipulate data transmitted between trusted devices. LE devices should be paired in a secure environment to minimize the risk of eavesdropping and MITM attacks. Just Works pairing should not be used.

	Security Issue or Vulnerability	Remarks
17	LE Security Mode 1 Level 1 does not require any security mechanisms (i.e., no authentication or encryption).	Similar to BR/EDR Security Mode 1, this is inherently insecure. LE Security Mode 1 Level 3 (authenticated pairing and encryption) is highly recommended instead.
All Versions		
18	Link keys can be stored improperly.	Link keys can be read or modified by an attacker if they are not securely stored and protected via access controls.
19	Strengths of the pseudo-random number generators (PRNG) are not known.	The Random Number Generator (RNG) may produce static or periodic numbers that may reduce the effective-ness of the security mechanisms. Bluetooth implementa-tions should use strong PRNGs based on NIST standards.
20	Encryption key length is negotiable.	The v3.0 and earlier specifications allow devices to negotiate encryption keys as small as one byte. Bluetooth LE requires a minimum key size of seven bytes. NIST strongly recommends using the full 128-bit key strength for both BR/EDR (E0) and LE (AES-CCM).
21	No user authentication exists.	Only device authentication is provided by the specification. Application-level security, including user authentication, can be added via overlay by the application developer.
22	End-to-end security is not performed.	Only individual links are encrypted and authenticated. Data is decrypted at intermediate points. End-to-end security on top of the Bluetooth stack can be provided by use of additional security controls.
23	Security services are limited.	Audit, non-repudiation, and other services are not part of the standard. If needed, these services can be incorpo-rated in an overlay fashion by the application developer.
24	Discoverable and/or connectable devices are prone to attack.	Any device that must go into discoverable or connectable mode to pair or connect should only do so for a minimal amount of time. A device should not be in discoverable or connectable mode all the time.

4.2. Bluetooth Threats

Bluetooth offers several benefits and advantages, but the benefits are not provided without risk. Bluetooth technology and associated devices are susceptible to general wireless networking threats, such as denial of service attacks, eavesdropping, MITM attacks, message modification, and resource misappropriation,[16] and are also threatened by more specific Bluetooth-related attacks, such as the following:

- **Bluesnarfing.** Bluesnarfing[17] enables attackers to gain access to a Bluetooth-enabled device by exploiting a firmware flaw in older devices. This attack forces a connection to a Bluetooth device, allowing access to data stored on the device including the device's international mobile equipment identity (IMEI). The IMEI is a unique identifier for each device that an attacker could potentially use to route all incoming calls from the user's device to the attacker's device.
- **Bluejacking.** Bluejacking is an attack conducted on Bluetooth-enabled mobile devices, such as cell phones. An attacker initiates bluejacking by sending unsolicited messages to the user of a Bluetooth-enabled device. The actual messages do not cause harm to the user's device, but they may entice the user to respond in some fashion or add the new contact to the device's address book. This message-sending attack resembles spam and phishing attacks conducted against e-mail users. Bluejacking can cause harm when a user initiates a response to a bluejacking message sent with a harmful intent.
- **Bluebugging.** Bluebugging[18] exploits a security flaw in the firmware of some older Bluetooth devices to gain access to the device and its commands. This attack uses the commands of the device without informing the user, allowing the attacker to access data, place phone calls, eavesdrop on phone calls, send messages, and exploit other services or features offered by the device.
- **Car Whisperer.** Car Whisperer[19] is a software tool developed by European security researchers that exploits a key implementation issue in hands-free Bluetooth car kits installed in automobiles. The Car Whisperer software allows an attacker to send to or receive audio from the car kit. An attacker could transmit audio to the car's speakers or receive audio (eavesdrop) from the microphone in the car.
- **Denial of Service.** Like other wireless technologies, Bluetooth is susceptible to DoS attacks. Impacts include making a device's Bluetooth interface unusable and draining the device's battery. These types of attacks are not significant and, because of the proximity required for Bluetooth use, can usually be easily averted by simply moving out of range.

- **Fuzzing Attacks.** Bluetooth fuzzing attacks consist of sending malformed or otherwise non-standard data to a device's Bluetooth radio and observing how the device reacts. If a device's operation is slowed or stopped by these attacks, a serious vulnerability potentially exists in the protocol stack.
- **Pairing Eavesdropping**. PIN/Legacy Pairing (Bluetooth 2.0 and earlier) and LE Pairing (Bluetooth 4.0) are susceptible to eavesdropping attacks. The successful eavesdropper who collects all pairing frames can determine the secret key(s) given sufficient time, which allows trusted device impersonation and active/passive data decryption.
- **Secure Simple Pairing Attacks.** A number of techniques can force a remote device to use Just Works SSP and then exploit its lack of MITM protection (e.g., the attack device claims that it has no input/output capabilities). Further, fixed passkeys could allow an attacker to perform MITM attacks as well.

4.3. Risk Mitigation and Countermeasures

Organizations should mitigate risks to their Bluetooth implementations by applying countermeasures to address specific threats and vulnerabilities. Some of these countermeasures cannot be achieved through security features built into the Bluetooth specifications.

The countermeasures recommended in the checklist in Section 4.4 do not guarantee a secure Bluetooth environment and cannot prevent all adversary penetrations. In addition, security comes at a cost—expenses related to security equipment, inconvenience, maintenance, and operation. Each organization should evaluate the acceptable level of risk based on numerous factors, which will affect the level of security implemented by that organization. To be effective, Bluetooth security should be incorporated throughout the entire lifecycle of Bluetooth solutions.[20]

FIPS Publication (PUB) 199 establishes three security categories—low, moderate, and high—based on the potential impact of a security breach involving a particular system. NIST SP 800-53 provides recommendations for minimum management, operational, and technical security controls for information systems based on the FIPS PUB 199 impact categories.[21]

The recommendations in NIST SP 800-53 should be helpful to organizations in identifying the controls needed to protect Bluetooth implementations in general, which should be used in addition to the specific recommendations for Bluetooth implementations listed in this document.

The first line of defense is to provide an adequate level of knowledge and understanding for those who will deal with Bluetooth-enabled devices. Organizations using Bluetooth technology should establish and document security policies that address the use of Bluetooth-enabled devices and users' responsibilities.

Organizations should include awareness-based education to support staff understanding and knowledge of Bluetooth. Policy documents should include a list of approved uses for Bluetooth and the type of information that may be transferred over Bluetooth networks. The security policy should also specify a proper password usage scheme. When feasible, a centralized security policy management approach should be used in coordination with an endpoint security product installed on the Bluetooth devices to ensure that the policy is locally and universally enforced.

The general nature and mobility of Bluetooth-enabled devices increases the difficulty of employing traditional security measures across the enterprise. Nevertheless, a number of countermeasures can be enacted to secure Bluetooth devices and communications, ranging from distance and power output to general operation practices. Several countermeasures that could be employed are provided in the checklist in Section 4.4.

4.4. Bluetooth Security Checklists

Table 4-2 provides a Bluetooth security checklist with guidelines and recommendations for creating and maintaining secure Bluetooth piconets.

For each recommendation or guideline in the checklist, a justification column lists areas of concern for Bluetooth devices, the security threats and vulnerabilities associated with those areas, risk mitigations for securing the devices from these threats, and vulnerabilities. In addition, for each recommendation three checklist columns are provided.

- The first column, the *Recommended Practice* column, if checked, means that this entry represents a recommendation for all organizations.
- The second column, the *Should Consider* column, if checked, means that the entry's recommendation should be considered carefully by an organization for one or more of the following reasons.

Table 4.2. Bluetooth Piconet Security Checklist

	Security Recommendation	Security Need, Requirement, or Justification	Checklist		
			Recommended Practice	Should Consider	Status
Management Recommendations					
1	Develop an organizational wireless security policy that addresses Bluetooth technology.	A security policy is the foundation for all other countermeasures.	√		
2	Ensure that Bluetooth users on the network are made aware of their security-related responsibilities regarding Bluetooth use.	A security awareness program helps users to follow practices that help prevent security incidents.	√		
3	Perform comprehensive security assessments at regular intervals to fully understand the organization's Bluetooth security posture.	Assessments help identify Bluetooth devices being used within the organization and help ensure the wireless security policy is being followed.	√		
4	Ensure that wireless devices and networks involving Bluetooth technology are fully understood from an architecture perspective and documented accordingly.	Bluetooth-enabled devices can contain various networking technologies and interfaces, allowing connections to local and wide area networks. An organization should understand the overall connectivity of each device to identify possible risks and vulnerabilities. These risks and vulnerabilities can then be addressed in the wireless security policy.	√		
5	Provide users with a list of precautionary measures they should take to better protect handheld Bluetooth devices from theft.	The organization and its employees are responsible for its wireless technology components because theft of those components could lead to malicious activities against the organization's information system resources.	√		
6	Maintain a complete inventory off all Bluetooth-enabled wireless devices and addresses (BD_ADDRs).	A complete inventory list of Bluetooth-enabled wireless devices can be referenced when conducting an audit that searches for unauthorized use of wireless technologies.		√	

Table 4.2. (Continued)

	Security Recommendation	Security Need, Requirement, or Justification	Checklist		
			Recommended Practice	Should Consider	Status
Technical Recommendations					
7	Change the default settings of the Bluetooth device to reflect the organization's security policy.	Because default settings are generally not secure, a careful review of those settings should be performed to ensure that they comply with the organizational security policy. For example, the default device name should usually be changed to be non-descriptive (i.e., so that it does not reveal the platform type).	√		
8	Set Bluetooth devices to the lowest necessary and sufficient power level so that transmissions remain within the secure perimeter of the organization.	Setting Bluetooth devices to the lowest necessary and sufficient power level ensures a secure range of access to authorized users. The use of Class 1 devices, as well as external amplifiers or high-gain antennas, should be avoided because of their extended range.	√		
9	Choose PIN codes that are sufficiently random, long and private. Avoid static and weak Pins, such as all zeroes.	PIN codes should be random so that malicious users cannot easily guess them. Longer PIN codes are more res-istant to brute force attacks. For Bluetooth v2.0 (or earlier) devices, an eight-character alphanumeric PIN should be used, if possible. The use of a fixed PIN is not acceptable.	√		
10	Ensure that link keys are not based on unit keys.	The use of shared unit keys can lead to successful spoofing, MITM, and eavesdropping attacks. The use of unit keys for security was deprecated in Bluetooth v1.2.	√		
11	For v2.1 and later devices using SSP, avoid using the "Just Works" association model. The device must verify that an authenticated link key was generated during pairing.	The "Just Works" association model does not provide MITM protection. Devices that only support Just Works (e.g., devices that have no input/output capability) should not be procured if similarly qualified devices that support one of the other association models (i.e., Numeric Comparison, OOB, or Passkey Entry) are available.	√		

	Security Recommendation	Security Need, Requirement, or Justification	Checklist		
			Recommended Practice	Should Consider	Status
12	For v2.1 and later devices using SSP, random and unique passkeys must be used for each pairing based on the Passkey Entry association model.	If a static passkey is used for multiple pairings, the MITM protection provided by the Passkey Entry association model is reduced.	√		
13	A Bluetooth v2.1 or later device using Security Mode 4 must fallback to Security Mode 3 for backward compatibility with v2.0and earlier devices (i.e., for devices that do not support Security Mode 4).	The Bluetooth specifications allow av2.1 device to fall back to any Security Mode for backward compatibility. This allows the option of falling back to Security Modes 1-3. As discussed earlier, Security Mode 3 provides the best security.	√		
14	LE devices and services should use Security Mode 1 Level 3whenever possible. LE Security Mode 1 Level 3 provides the highest security available for LE devices	Other LE security modes allow unauthenticated pairing and/or no encryption.	√		
15	Unneeded and unapproved service and profiles should bedisabled.[22]	Many Bluetooth stacks are designed to support multiple profiles and associated services. The Bluetooth stack on a device should be locked down to ensure only required and approved profiles and services are available for use.	√		
16	Bluetooth devices should be configured by default as undiscoverable and remain undiscoverable except as needed for pairing.	This prevents visibility to other Bluetooth devices except when discovery is absolutely required. In addition, the default Bluetooth device names sent during discovery should be changed to non-identifying values.	√		
17	Invoke link encryption for all Bluetooth connections.	Link encryption should be used to secure all data transmissions during a Bluetooth connection; otherwise, transmitted data are vulnerable to eavesdropping.	√		
18	If multi-hop wireless communication is being used, ensure that encryption is enabled on every link in the communication chain.	One unsecured link results in compromising the entire communication chain.	√		

Table 4.2. (Continued)

	Security Recommendation	Security Need, Requirement, or Justification	Checklist		
			Recommended Practice	Should Consider	Status
19	Ensure device mutual authentication is performed for all connections.	Mutual authentication is required to provide verification that all devices on the network are legitimate.	√		
20	Enable encryption for all broadcast transmissions (Encryption Mode 3).	Broadcast transmissions secured by link encryption provide a layer of security that protects these transmissi-ons from user interception for malicious purposes.	√		
21	Configure encryption key sizes to the maximum allowable (128-bit).	Using maximum allowable key sizes provides protection from brute force attacks.	√		
22	Use application-level authentication and encryption atop the Bluetooth stack for sensitive data communication.	Bluetooth devices can access link keys from memory and automatically connect with previously paired devices. Incorporating application-level software that implements authentication and encryption will add an extra layer of security. Passwords and other authentication mechanisms, such as biometrics and smart cards, can be used to provide user authentication for Bluetooth devices. Employing higher layer encryption (particularly FIPS140 validated) over the native encryption will further protect the data in transit.		√	
23	Deploy user authentication overlays such as biometrics, smart cards, two-factor authentication, or public key infrastructure (PKI).	Implementing strong authentication mechanisms can minimize the vulnerabilities associated with passwords and PINs.		√	
Operational Recommendations					
24	Ensure that Bluetooth capabilities are disabled when they are not in use.	Bluetooth capabilities should be disabled on all Bluetooth devices, except when the user explicitly enables Bluetooth to establish a connection. This minimizes exposure to potential malicious activities.	√		

	Security Recommendation	Security Need, Requirement, or Justification	Checklist		
			Recommended Practice	Should Consider	Status
		For devices that do not support disabling Bluetooth (e.g., headsets), the entire device should be shut off when not in use.			
25	Perform pairing as inferquently as possible, ideally in a secure area where attackers cannot realistically observe the passkey entry and intercept Bluetooth pairing messages. (Note: A" secure area" is defined as a non-public area that is indoors away from windows in locations with physical access controls.)Users should not respond to any messages requesting a PIN, unless the user has initiated a pairing and is certain the PIN request is being sent by one of the user's devices.[23]	Pairing is a vital security function and requires that users maintain a security awareness of possible eavesdroppers. If an attacker can capture the transmitted frames associated with pairing, determining the link key is straightforward forpre-v2.1 and v4.0 devices since security is solely dependent on PIN entropy and length. This recommendation also applies tov2.1/3.0 devices, although similar eavesdropping attacks against SSP have not yet been documented.	√		
26	A BR/EDR service-level security mode (i.e., Security Mode 2 or 4) should only be used in a controlled and well-understood environment.	Security Mode 3 provides link-level security prior to link establishment, while Security Modes 2 and 4 allow link-level connections before any authentication or encryption is established. NIST highly recommends that devices use Security Mode 3.	√		
27	Ensure that portable devices with Bluetooth interfaces are configured with a password.	This helps prevent unauthorized access if the device is lost or stolen.	√		
28	In the event a Bluetooth device is lost or stolen, users should immediately delete the missing device from the paired device lists of all other Bluetooth devices.	This policy will prevent an attacker from using the lost or stolen device to access another Bluetooth device owned by the user(s).	√		
29	Install antivirus software on Bluetooth-enabled hosts that support such host-based security software.	Antivirus software should be installed to ensure that known malware is not introduced to the Bluetooth network.	√		

Table 4.2. (Continued)

	Security Recommendation	Security Need, Requirement, or Justification	Checklist		
			Recommended Practice	Should Consider	Status
30	Fully test and regularly deploy Bluetooth software and firmware patches and upgrades.	Newly discovered security vulnerabilities of vendor products should be patched to prevent malicious and inadvertent exploits. Patches should be fully tested before implementation to confirm that they are effective.	√		
31	Users should not accept transmissions of any kind from unknown or suspicious devices.	With the increase in the number of Bluetooth-enabled devices, it is important that users only establish	√		
	These types of transmissions include messages, files, and images.	connections with other trusted devices and only accept content from these trusted devices			
32	Fully understand the impacts of deploying any security feature or product prior to deployment	To ensure a successful deployment, an organization should fully understand the technical, security, operational, and personnel requirements prior to implementation.	√		
33	Designate an individual to track the progress of Bluetooth security products and standards (perhaps via the Bluetooth SIG) and the threats and vulnerabilities with the technology.	An individual designated to track the latest technology enhancements, standards (perhaps via Bluetooth SIG), and risks will help to ensure the continued secure use of Bluetooth.		√	

- First, implementing the recommendation may provide a higher level of security for the wireless environment by offering some additional protection.
- Second, the recommendation supports a defense-in-depth strategy.
- Third, it may have significant performance, operational, or cost impacts. In summary, if the *Should Consider* column is checked, organizations should carefully consider the option and weigh the costs versus the benefits.

The last column, *Status*, is intentionally left blank to allow organization representatives to use this table as a true checklist. For instance, an individual performing a wireless security audit in a Bluetooth environment can quickly check off each recommendation for the organization, asking, "Have I done this?"

APPENDIX A. GLOSSARY OF TERMS

Selected terms used in the publication are defined below.

Access Point (AP): A device that logically connects wireless client devices operating in infrastructure to one another and provides access to a distribution system, if connected, which is typically an organization's enterprise wired network.

Ad Hoc Network: A wireless network that dynamically connects wireless client devices to each other without the use of an infrastructure device, such as an access point or a base station.

Claimant: The Bluetooth device attempting to prove its identity to the verifier during the Bluetooth connection process.

Infrastructure Network: A wireless network that requires the use of an infrastructure device, such as an access point or a base station, to facilitate communication between client devices.

Media Access Control (MAC): A unique 48-bit value that is assigned to a particular wireless network interface by the manufacturer.

Piconet: A small Bluetooth network created on an ad hoc basis that includes two or more devices.

Range: The maximum possible distance for communicating with a wireless network infrastructure or wireless client.

Scatternet: A chain of piconets created by allowing one or more Bluetooth devices to each be a slave in one piconet and act as the master for another piconet simultaneously. A scatternet allows several devices to be networked over an extended distance.

Verifier: The Bluetooth device that validates the identity of the claimant during the Bluetooth connection process.

Wireless Local Area Network (WLAN): A group of wireless access points and associated infrastructure within a limited geographic area, such as an office building or building campus, that is capable of radio communications. WLANs are usually implemented as extensions of existing wired LANs to provide enhanced user mobility.

Wireless Personal Area Network (WPAN): A small-scale wireless network that requires little or no infrastructure and operates within a short range. A WPAN is typically used by a few devices in a single room instead of connecting the devices with cables.

APPENDIX B. ACRONYMS AND ABBREVIATIONS

Selected acronyms and abbreviations used in the publication are defined below.

8DPSK	8 phase Differential Phase Shift Keying
AC	Alternating Current
ACO	Authenticated Cipher Offset
AES-CCM	Advanced Encryption Standard–Counter with CBC-MAC
AFH	Adaptive Frequency Hopping
AMP	Alternate MAC/PHY
AP	Access Point

ATT	Attribute Protocol
BR	Basic Rate
CSRK	Connection Signature Resolving Key
CTIA	Cellular Telecommunications and Internet Association
dBm	Decibels referenced to one milliwatt
DHK	Diversifier Hiding Key
DISA	Defense Information Systems Agency
DIV	Diversifier
DoD	Department of Defense
DoS	Denial of Service
DQPSK	Differential Quaternary Phase Shift Keying
ECDH	Elliptic Curve Diffie-Hellman
EDIV	Encrypted Diversifier
EDR	Enhanced Data Rate
ER	Encryption Root
FHSS	Frequency Hopping Spread Spectrum
FIPS	Federal Information Processing Standard
FISMA	Federal Information Security Management Act
GAP	Generic Access Profile
GATT	Generic Attribute Protocol
GFSK	Gaussian Frequency-Shift Keying
GHz	Gigahertz
HCI	Host Controller Interface
HS	High Speed
IBC	Iterated Block Cipher
IEEE	Institute of Electrical and Electronics Engineers
IMEI	International Mobile Equipment Identity
IRK	Identity Resolving Key
ISM	Industrial, Scientific, and Medical
ITL	Information Technology Laboratory
kbps	Kilobits per second
KG	Key Generator
KSG	Key Stream Generator
L2CAP	Logical Link Control and Adaptation Protocol
LAN	Local Area Network
LE	Low Energy
LFSR	Linear Feedback Shift Register
LTK	Long-Term Key
m	Meter

MAC	Medium Access Control
Mbps	Megabits per second
MHz	Megahertz
MITM	Man-in-the-Middle
mW	Milliwatt
NFC	Near Field Communication
NIST	National Institute of Standards and Technology
NVD	National Vulnerability Database
OFDM	Orthogonal Frequency-Division Multiplexing
OMB	Office of Management and Budget
OOB	Out of Band
P2P	Peer-to-Peer
PAL	Protocol Adaptation Layer
PC	Personal Computer
PDA	Personal Digital Assistant
PHY	Physical Layer
PIN	Personal Identification Number
PKI	Public Key Infrastructure
PRNG	Pseudo-Random Number Generator
PUB	Publication
RF	Radio Frequency
RNG	Random Number Generator
RSSI	Received Signal Strength Indication
SAFER	Secure And Fast Encryption Routine
SDP	Service Discovery Protocol
SIG	Special Interest Group
SMP	Security Manager Protocol
SP	Special Publication
SRES	Signed Response
SSP	Secure Simple Pairing
STK	Short Term Key
TDM	Time Division Multiplexing
TK	Temporary Key
USB	Universal Serial Bus
UWB	Ultra Wideband
WLAN	Wireless Local Area Network
WPAN	Wireless Personal Area Network

APPENDIX C. REFERENCES

The list below provides references for the publication.

[1] Bluetooth Special Interest Group, Bluetooth specifications. https://www.bluetooth.org/Technical/Specifications/adopted.htm

[2] Bluetooth Special Interest Group, "Bluetooth Security White Paper", May 2002. http://grouper.ieee.org/groups/1451/5/ Comparison% 20of% 20PHY/Bluetooth_24Security_Paper.pdf

[3] Defense Information Systems Agency, "DoD Bluetooth Peripheral Device Security Requirements", 16 July 2010. http://iase.disa.mil/ stigs/downloads/pdf/dod_bluetooth_requirements_spec_20100716.pdf

[4] Gehrmann, C. Persson, J. & Smeets, B. (2004). *Bluetooth Security*, Artech House.

[5] Meier, Y. Lu, W. & Vaudenay, S. (2005). "*The Conditional Correlation Attack: A Practical Attack on Bluetooth Encryption*", In Advances of Cryptology, CRYPTO 2005 vol. 3621, pages 97–117, August. http://lasecwww.epfl.ch/pub/ lasec/doc/LMV05.pdf

[6] National Security Agency, "Bluetooth Security Factsheet", December 2007. http://www.nsa.gov/ia/_files/factsheets/I732-016R-07.pdf

[7] Shaked Y. & Wool, A. (2005). "Cracking the Bluetooth PIN", In *Proc. 3rd USENIX/ACM Conf. Mobile Systems, Applications, and Services (MobiSys)*, pages 39–50, Seattle, WA, June. http://www.usenix.org/event/mobisys05/tech/full_papers/ shaked/ shaked.pdf

APPENDIX D. RESOURCES

The lists below provide examples of resources related to Bluetooth technologies that may be helpful to readers.

Documents

Name	URL
Bluetooth SIG Specifications	https://www.bluetooth.org/Technical/Specifications/adopted.htm
FIPS 140-2, *Security Requirements for Cryptographic Modules*	http://csrc.nist.gov/publications/fips/fips 140-2/fips1402.pdf

Name	URL
FIPS 180-4, *Secure Hash Standard (SHS)*	http://csrc.nist.gov/publications/fips/fips180-4/fips-180-4.pdf
FIPS 197, *Advanced Encryption Standard*	http://csrc.nist.gov/publications/fips/fips197/fips-197.pdf
FIPS 199, *Standards for Security Categorization of Federal Information and Information Systems*	http://csrc.nist.gov/publications/fips/fips199/FIPS-PUB-199-final.pdf
GAO-05-383, *Information Security: Federal Agencies Need to Improve Controls over Wireless Networks*	http://www.gao.gov/new.items/d05383.pdf
NIST SP 800-37Revision 1, *Guide for Applying the Risk Management Framework to Federal Information Systems: A Security Life Cycle Approach*	http://csrc.nist.gov/publications/PubsSPs.html#800-37
NIST SP 800-53 Revision 3, *Recommended Security Controls for Federal Information Systems and Organizations*	http://csrc.nist.gov/publications/PubsSPs.html#800-53
NIST SP 800-63 Revision 1, *Electronic Authentication Guideline*	http://csrc.nist.gov/publications/PubsSPs.html#800-63-Rev1
NIST SP 800-64 Revision 2, *Security Considerations in the Information System Development Life Cycle*	http://csrc.nist.gov/publications/PubsSPs.html#800-64
NIST SP 800-70 Revision 2, *National Checklist Program for IT Products—Guidelines for Checklists Users and Developers*	http://csrc.nist.gov/publications/PubsSPs.html#800-70
NIST SP 800-114, *User's Guide to Securing External Devices for Telework and Remote Access*	http://csrc.nist.gov/publications/PubsSPs.html#800-114

Resource Sites

Name	URL
Bluetooth Special Interest Group	http://www.bluetooth.com/, https://www.bluetooth.org/
Cellular Telecommunications and Internet Association (CTIA)	http://www.ctia.org/
FIPS-Validated Cryptographic Modules	http://csrc.nist.gov/groups/ STM/
IEEE 802.15 Working Group for Wireless Personal Area Networks	http://www.ieee802.org/15/

Name	URL
NIST National Vulnerability Database (NVD)	http://nvd.nist.gov/
NIST's National Checklist Program	http://checklists.nist.gov/
Trifinite Group (Bluetooth Security Research)	http://trifinite.org/
Wireless Vulnerabilities and Exploits	http://www.wve.org/

End Notes

[1] As discussed in Section 2.2, the term "piconet" applies to both ad hoc and infrastructure Bluetooth networks.

[2] The Bluetooth SIG website (http://www.bluetooth.com/) is a resource for Bluetooth-related information and provides numerous links to other sources of information.

[3] For more information, see the IEEE website at http://grouper.ieee.org/groups/802/15/.

[4] For more information, see Dominic Spill and Andrea Bittau's 2007 research paper: http://www.usenix.org/event/woot07/tech/full_papers/spill/spill.pdf

[5] The ranges listed in Table 2-1 are the designed operating ranges. Attackers may be able to intercept communications at significantly larger distances, especially if they use high-gain antennas and high-sensitivity receivers.

[6] Having a remote device's clock information is not needed to make a connection, but it will speed up the connection process.

[7] Note that a particular device can only be the master of one piconet at any given time.

[8] The Bluetooth Security White Paper from the Bluetooth Special Interest Group is available at http://grouper.ieee.org/groups/1451/5/Comparison%20of%20PHY/Bluetooth_24Security_Paper.pdf.

[9] This information is derived from the Bluetooth 2.1 specification, which is available at https://www.bluetooth.org/Technical/Specifications/adopted.htm.

[10] Download a copy of NIST SP 800-97 here: http://csrc.nist.gov/publications/nistpubs/800-97/SP800-97.pdf

[11] The E_1 authentication function is based on the SAFER+ algorithm. SAFER stands for Secure And Fast Encryption Routine. The SAFER algorithms are iterated block ciphers (IBCs). In an IBC, the same cryptographic function is applied for a specified number of rounds.

[12] LFSRs are used in coding (error control coding) theory and cryptography. LFSR-based key stream generators (KSG), composed of exclusive-OR gates and shift registers, are common in stream ciphers and are very fast in hardware.

[13] Y. Lu, W. Meier, and S. Vaudenay. "The Conditional Correlation Attack: A Practical Attack on Bluetooth Encryption." http://lasecwww.epfl.ch/pub/lasec/doc/LMV05.pdf

[14] http://csrc.nist.gov/publications/nistpubs/800-108/sp800-108.pdf

[15] DIV = Y XOR EDIV where Y = dm(DHK, rand) and DHK is the Diversifier Hiding Key.

[16] Additional information on general wireless security threats is available in Section 3 of NIST SP 800-48 Revision 1, *Guide to Securing Legacy IEEE 802.11 Wireless Networks* (http://csrc.nist.gov/publications/nistpubs/800-48-rev1/SP800-48r1.pdf).

[17] http://trifinite.org/trifinite_stuff_bluesnarf.html

[18] http://trifinite.org/trifinite_stuff_bluebug.html

[19] http://trifinite.org/trifinite_stuff_carwhisperer.html

[20] For more information about technology lifecycles, see NIST SP 800-64, *Security Considerations in the Information System Development Life Cycle* (http://csrc.nist.gov/publications/PubsSPs.html#800-64).

[21] FIPS PUB 199, *Standards for Security Categorization of Federal Information and Information Systems*, is available at http://csrc.nist.gov/publications/fips/fips199/FIPS-PUB-199-final.pdf. NIST SP 800-53 Revision 3, *Recommended Security Controls for Federal Information Systems and Organizations*, is available at http://csrc.nist.gov/publications/PubsSPs.html#800-53.

[22] Derived from requirement 1.4 in the DoD Bluetooth Peripheral Security Requirements (16 July 2010), available at http://iase.disa.mil/stigs/downloads/pdf/dod_bluetooth_requirements_spec_20100716.pdf

[23] Derived from requirement 4.1.5 in the DoD Bluetooth Peripheral Security Requirements (16 July 2010), available at http://iase.disa.mil/stigs/downloads/pdf/dod_bluetooth_requirements_spec_20100716.pdf

In: Bluetooth and Wireless Local Area Networks ISBN: 978-1-62417-259-5
Editor: Rene I. Jenkins

Chapter 2

GUIDELINES FOR SECURING WIRELESS LOCAL AREA NETWORKS (WLANS)*

Murugiah Souppaya and Karen Scarfone

EXECUTIVE SUMMARY

A *wireless local area network (WLAN)* is a group of wireless networking devices within a limited geographic area, such as an office building, that exchange data through radio communications. WLAN technologies are based on the IEEE 802.11 standard and its amendments.[1] The fundamental components of an IEEE 802.11 WLAN (hereafter referred to as a "WLAN" in this publication) are client devices, such as laptops and smartphones, and access points (APs), which logically connect client devices with a distribution system, typically the organization's wired network infrastructure. Some WLANs also use wireless switches, which act as intermediaries between APs and the distribution system.

The security of each WLAN is heavily dependent on how well each WLAN component—including client devices, APs, and wireless switches—is secured throughout the WLAN lifecycle, from initial WLAN design and deployment through ongoing maintenance and monitoring. Unfortunately, WLANs are typically less secure than their wired counterparts for several reasons, including the ease of access to the WLAN and the weak security

---

* This is an edited, reformatted and augmented version of the NIST Publication 800-153, U.S. Department of Commerce, dated February 2012.

configurations often used for WLANs (to favor convenience over security). The purpose of this publication is to help organizations improve their WLAN security by providing recommendations for WLAN security configuration and monitoring. This publication supplements other NIST publications by consolidating and strengthening their key recommendations.

Organizations should implement the following guidelines to improve the security of their WLANs.

Have standardized security configurations for common WLAN components, such as client devices and APs.

A standardized configuration provides a base level of security, reducing vulnerabilities and lessening the impact of successful attacks. Standardized configurations can also significantly reduce the time and effort needed to secure WLAN components and verify their security, particularly if the configuration can be deployed and verified through automated means.

When planning WLAN security, consider the security not only of the WLAN itself, but also how it may affect the security of other networks.

A WLAN is usually connected to an organization's wired networks, and WLANs may also be connected to each other. For WLANs that need wired network access, their client devices should be allowed access only to the necessary hosts on the wired network using only the required protocols. Also, an organization should have separate WLANs if there is more than one security profile for WLAN usage; for example, an organization should have logically separated WLANs for external use (such as guests) and internal use. Devices on one WLAN should not be able to connect to devices on a logically separated WLAN.

Have policies that clearly state which forms of dual connections are permitted or prohibited for WLAN client devices, and enforce these policies through the appropriate security controls.

The term "dual connected" generally refers to a client device that is connected to both a wired network and a WLAN at the same time. If an attacker gains unauthorized wireless access to a dual-connected client device, the attacker could then use it to access or attack resources on the wired network. Organizations should consider the risks posed not only by the traditional form of dual connectness, but also by other forms involving multiple wireless networks. It is common today for client devices to be connected to multiple wireless networks simultaneously, such as cell phone, WiMAX, Bluetooth, and WLAN networks. Organizations should assess the risk of the possible combinations of network technologies for their WLAN client devices and determine how those risks should be mitigated. If one or more of the networks cannot have its risk mitigated to an acceptable level, then dual connections involving that network may pose too much risk to the organization and may need to be prohibited.

Ensure that the organization's WLAN client devices and APs have configurations at all times that are compliant with the organization's WLAN policies.

After designing WLAN security configurations for client devices and APs, an organization should determine how the configurations will be implemented, evaluate the effectiveness of the implementations, deploy the implementations to the appropriate devices, and maintain the configurations and their implementations throughout the devices' lifecycles. Organizations should standardize, automate, and centralize as much of their WLAN security configuration implementation and maintenance as practical. This allows organizations to implement consistent WLAN security throughout the enterprise, to detect and correct unauthorized changes to configurations, and to react quickly when newly identified vulnerabilities or recent incidents indicate a need to change the WLAN's security configuration.

Perform both attack monitoring and vulnerability monitoring to support WLAN security.

Security monitoring is important for all systems and networks, but it is generally even more important for WLANs because of the increased risks that they face. Organizations should continuously monitor their WLANs for both WLAN-specific and general (wired network) attacks.

Organizations should do largely the same vulnerability monitoring for WLAN components that they do for any other software: identifying patches and applying them, and verifying security configuration settings and adjusting them as needed. These actions should be performed at least as often for WLAN components as they are for the organization's equivalent wired systems.

Conduct regular periodic technical security assessments for the organization's WLANs.

These assessments should be performed at least annually to evaluate the overall security of the WLAN. In addition, organizations should perform periodic assessments at least quarterly unless continuous monitoring of WLAN security is already collecting all of the necessary information about WLAN attacks and vulnerabilities needed for assessment purposes.

1. Introduction

1.1. Authority

The National Institute of Standards and Technology (NIST) developed this document in furtherance of its statutory responsibilities under the Federal Information Security Management Act (FISMA) of 2002, Public Law 107-347.

NIST is responsible for developing standards and guidelines, including minimum requirements, for providing adequate information security for all agency operations and assets; but such standards and guidelines shall not apply to national security systems. This guideline is consistent with the requirements of the Office of Management and Budget (OMB) Circular A-130, Section 8b(3), "Securing Agency Information Systems," as analyzed in A-130, Appendix IV: Analysis of Key Sections. Supplemental information is provided in A-130, Appendix III. This guideline has been prepared for use by Federal agencies. It may be used by nongovernmental organizations on a voluntary basis and is not subject to copyright, though attribution is desired.

Nothing in this document should be taken to contradict standards and guidelines made mandatory and binding on Federal agencies by the Secretary of Commerce under statutory authority, nor should these guidelines be interpreted as altering or superseding the existing authorities of the Secretary of Commerce, Director of the OMB, or any other Federal official.

1.2. Purpose and Scope

The purpose of this publication is to provide organizations with recommendations for improving the security configuration and monitoring of their IEEE 802.11 wireless local area networks (WLANs) and their devices connecting to those networks. The scope of this publication is limited to unclassified wireless networks and unclassified facilities within range of unclassified wireless networks.

This publication supplements other NIST publications by consolidating and strengthening their key recommendations, and it points readers to the appropriate NIST publications for additional information (see Appendix C for the full list of references and Appendix A for a list of major security controls relevant for WLAN security). This publication does not eliminate the need to follow recommendations in other NIST publications, such as [SP800-48] and [SP800-97]. If there is a conflict between recommendations in this publication and another NIST wireless publication, the recommendation in this publication takes precedence.

1.3. Audience

The primary audience for this publication is security professionals, network professionals, system administrators, and others who are responsible for planning, implementing, maintaining, and monitoring the security of their organization's WLANs and the devices that connect to those WLANs.

1.4. Document Structure

The remainder of this document is composed of the following sections and appendices:

- Section 2 provides recommendations for WLAN security configuration, including configuration design, implementation, evaluation, and maintenance.
- Section 3 presents an overview of WLAN security monitoring and gives related recommendations, including criteria for selecting monitoring tools and guidelines for determining how often to perform monitoring
- Appendix A lists the major controls from NIST Special Publication 800-53, *Recommended Security Controls for Federal Information Systems and Organizations* that affect WLAN security.

- Appendix B provides a list of acronyms and abbreviations used in this document.
- Appendix C lists references for this document.

2. WLAN Security Configuration

Wireless networking enables computing devices with wireless capabilities to use computing resources without being physically connected to a network. The devices simply need to be within a certain distance (known as the range) of the wireless network infrastructure. Wireless local area networks (WLANs) are groups of wireless networking devices within a limited geographic area, such as an office building, that are capable of exchanging data through radio communications. WLANs are usually implemented as extensions to existing wired local area networks (LANs) to provide enhanced user mobility and network access. WLAN technologies are based on the IEEE 802.11 standard and its amendments. Throughout the rest of this publication, the generic term "WLAN" refers to an IEEE 802.11 WLAN.

The two fundamental types of WLAN components are client devices (such as laptops and smartphones) and access points (APs), which logically connect client devices with a distribution system (DS), typically the organization's wired network infrastructure. The DS is the means by which client devices can communicate with the organization's wired LANs and external networks such as the Internet. Some WLANs also use wireless switches, which act as intermediaries between APs and the DS. The purpose of the switch is to assist administrators in managing the WLAN infrastructure. Figure 1 shows a simplified view of WLAN components that includes a wireless switch. WLANs without wireless switches have a similar architecture, except that the APs connect directly to the DS.

The security of each of the WLAN components—including client devices, APs, and wireless switches—is heavily dependent on their WLAN security configuration. This section explains why having standardized security configurations is important for WLAN components and provides recommendations for designing, implementing, evaluating, and maintaining those configurations, particularly for client devices. As explained in Section 1.2, the recommendations presented in this section supplement those provided for specific WLAN technologies by other NIST publications [SP800-48, SP800-97].

This section does not provide an exhaustive explanation of the entire security configuration lifecycle; rather, it highlights a few topics of particular relevance to WLAN security. Section 2.1 discusses security configuration design, while Section 2.2 focuses on security configuration implementation, evaluation, and maintenance.

2.1. Configuration Design

Organizations should have standardized security configurations for their common WLAN components, such as client devices and APs. A standardized configuration provides a base level of security, reducing vulnerabilities and lessening the impact of successful attacks.

Standardized configuration use improves the consistency and predictability of security, in conjunction with user training and awareness activities and other supporting security controls. Standardized configurations can also provide a large resource savings by reducing the time needed to secure each WLAN device and to verify its configuration for security assessments, audits, etc., particularly if the configuration can be deployed and verified through automated means.

This section focuses on two noteworthy aspects of configuration design: gathering needs and designing WLAN architectures.

2.1.1. Needs Gathering

Before designing a WLAN security architecture or WLAN component security configurations, an organization should gather information on needs, particularly operational and security related ones. This should include identifying relevant WLAN security requirements from applicable laws, policies, regulations, etc. For Federal agencies, this often includes requirements from OMB, the Government Accountability Office (GAO), the Department of Homeland Security (DHS), and other agencies. Another part of needs gathering is identifying and reviewing recommended WLAN security practices from Federal agencies (e.g., NIST Special Publications, DISA Security Technical Implementation Guides), WLAN vendors, and other parties [NCP]. See Section 9 of [SP800-94] for examples of possible requirements to include in needs gathering.

In addition to identifying these requirements and recommendations, organizations should also determine what threats their WLAN security faces. Organizations should conduct risk assessments to identify the threats against their WLANs and determine the effectiveness of existing security controls in counteracting the threats; they then should perform risk mitigation to decide what additional measures (if any) should be implemented, as discussed in [SP800-37]. Performing risk assessments and mitigation helps organizations decide how their WLANs should be secured. See Section 3.1 for an overview of common WLAN threats.

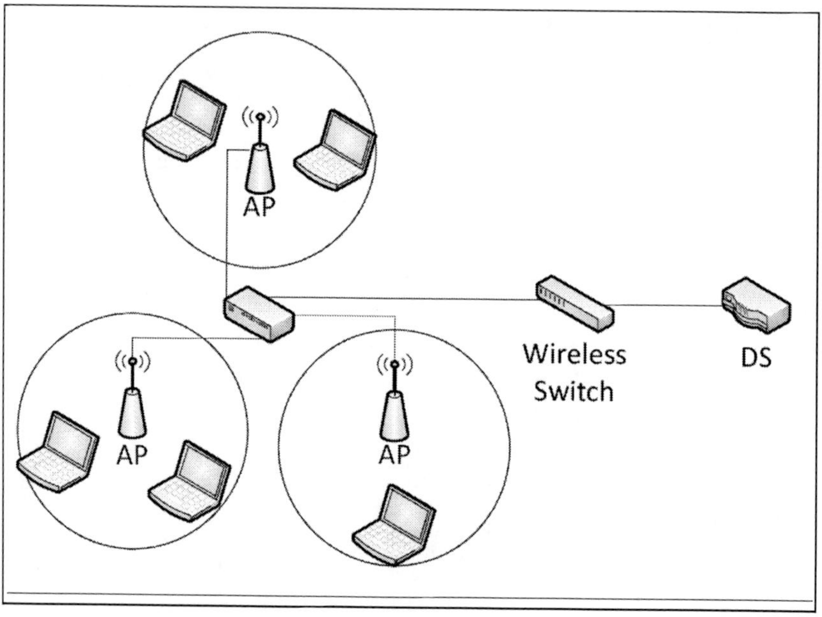

Figure 1. Simplified View of WLAN Architecture.

2.1.2. WLAN Architecture

When planning WLAN security, configuration designers should consider the security not only of the WLAN itself, but also how it may affect other networks that are accessible through it, such as internal wired networks reachable from the WLAN. An important principle of WLAN security is to separate WLANs with different security profiles. For example, there should be separate WLANs for external (guest, etc.) and internal use. Devices on an organization's external WLAN should not be able to connect through that WLAN to devices on another of the organization's WLANs. This helps to protect the organization's other networks and devices from external devices and users. Organizations often set up external WLANs primarily to provide Internet access to visitors; such WLANs should be architected so that their traffic does not traverse the organization's internal networks. For external WLANs that do need internal network access, WLAN client devices should be allowed access only to the necessary hosts or subnets using only the required protocols.

Another architectural issue mentioned in the WLAN reference architecture document and discussed in more detail in [GAO-11-43] is dual connected client devices. The term "dual connected" generally refers to a device that is connected to both a wired network and a WLAN at the same time. The primary concern with dual connected configurations is that an attacker may be able to gain unauthorized wireless access to the client device and then use it to attack resources on the wired network. Essentially this is allowing an attacker to exploit a lower-security network in order to gain access to a higher-security network. One possible scenario is an attacker tunneling traffic from the higher-security network to the lower-security network through the client device instead of following the intended network architecture, and thus avoiding network-based security controls intended for the higher-security network. Dual connected configurations also generally violate the principle of disabling unneeded network services to reduce attack surface; if the device is already connected to a wired network access, WLAN access is usually redundant.

Organizations should not only consider simultaneous wired network and WLAN use, but other forms of dual connectedness involving their WLAN client devices. With the increasing variety and popularity of wireless networking technologies, it is common today for devices to be connected to multiple wireless networks simultaneously. For example, most smartphones can use cell phone networks, WLANs, and Bluetooth networks simultaneously, while they are also connected to wired laptops/desktops (and possibly their wired networks) through a cabled connection (e.g., USB). It is also increasingly common for laptops to have multiple wireless interfaces, such as both WLAN and WiMAX interfaces, or to be configured to accept removable media-based WLAN interfaces. A single laptop with multiple WLAN interfaces could have simultaneous connections to multiple WLANs, such as an organization WLAN and an external WLAN.

Organizations should assess the risk of the possible combinations of network technologies for their WLAN client devices and determine how those risks should be mitigated. This does not mean that all forms of dual connectedness should automatically be prohibited; examples of use cases that are often permitted include a smartphone attaching to both a WLAN and a Bluetooth-networked earbud simultaneously, and a laptop attaching to both a WLAN and a Bluetooth-networked keyboard and mouse simultaneously.

However, the security of such use cases is largely dependent on the security of all of the networks. If one or more of the networks cannot have its risk mitigated to an acceptable level, then dual connections involving that network may pose too much risk to the organization and may need to be prohibited. The primary risk-related issue to be considered is the likelihood of an attacker accessing and manipulating legitimate communications and the possible and typical impact of such an attack.

Organizations should have policies that clearly state which forms of dual connections are permitted or prohibited for their WLAN client devices under various circumstances. Organizations should enforce these policies through the appropriate security controls, including the actions listed below:

- For all their WLAN client devices: disable all network interfaces that are not authorized for any use (including during contingency plans for business continuity, disaster recovery, etc.), and configure the device so that the user cannot enable them or otherwise circumvent the restrictions.
- For all their WLAN client devices not authorized for dual connections:
  - Implement the appropriate technical security controls (discussed below the bullets) so that all dual connected configurations are prohibited.
  - If feasible, configure the devices to disable bridging (passing traffic between the networks). This is precautionary in case an unauthorized dual connection occurs.
- For all their WLAN client devices authorized for dual connections:
  - Implement the appropriate technical security controls (discussed below the bullets) so that the authorized dual connected configurations are only active when necessary and that all other dual connected configurations are prohibited.
  - Configure the devices to disable bridging (passing traffic between the networks) unless absolutely necessary.

There are several options for implementing WLAN client device dual connection policies, such as prohibiting unauthorized configurations from being used and disabling authorized dual connections when not needed. To enforce such restrictions, organizations should rely on automated technical controls whenever feasible. Non-technical controls are usually not effective enough; for example, it is generally not feasible to rely on users to remember to always promptly disable WLAN interfaces every time they are no longer needed. Each organization should evaluate the possible controls for implementing dual connection policies on their WLAN client devices and then choose the combinations of controls that are most appropriate, providing sufficient security while also permitting necessary functionality.

Some of the possible controls are preventative (implementing and enforcing a configuration), while others are detective (monitoring a configuration, monitoring network activity, alerting when a problem is detected, etc.) Preventative controls are generally preferable to detective controls, but it is even stronger to use both preventative and detective controls together. See Section 3 for more information on monitoring (detective) controls.

Preventative controls may need to enforce granular policies. For example, an organization might permit only wired network usage while at their headquarters, but might permit wired or wireless access to external networks. Preventative controls may also need to provide flexibility; organizations often need to allow users to attach the organization's mobile client devices to new external WLANs, and thus the users need some ability to manage their devices' WLAN configuration.

Examples of preventative controls include the following:

- Configure the device's BIOS so that WLAN connections are automatically terminated when a wired connection is detected. The BIOS setting for this is often called LAN/WLAN switching.
- Enable specialized software-based controls that permit either WLAN or wired network access, but not both simultaneously. These controls could be built into the operating system (OS), provided as part of the WLAN driver or management software, provided by the device manufacturer (e.g., laptop vendor), or acquired from third parties. These controls typically favor wired connections over WLAN because of their relative reliability, performance, and security.
- Configure host-based network security tools (e.g., host-based firewalls, host-based intrusion detection and prevention systems) to prevent multiple network interfaces from being used at one time.
- Specify and enforce authorized network profiles and/or unauthorized profiles through OS/domain controls, third party policy-based software, etc.

There are other technical WLAN security controls, such as configuring clients so that they will not automatically connect to any WLANs they detect, that are beneficial for WLAN security but not robust enough to prevent dual connections.

2.2. Configuration Implementation, Evaluation, and Maintenance

After designing a WLAN security configuration, an organization should determine how the configuration will be implemented, evaluate the effectiveness of the implementation, deploy the implementation to the appropriate devices, and maintain the configuration and its implementation throughout the devices' lifecycles. Organizations should ensure that their WLAN client devices and APs have configurations at all times that are compliant with the organization's WLAN policies.

Organizations should standardize, automate, and centralize as much of their WLAN security configuration implementation and maintenance as practical, particularly for their WLAN client devices and access points. This allows organizations to implement consistent WLAN security throughout the enterprise, to detect and correct unauthorized changes to configurations, and to react quickly when newly identified vulnerabilities or recent incidents indicate a need to change the WLAN's security configuration.

Organizations can often leverage existing configuration management solutions for their WLAN security configuration management. Many organizations already have general configuration management solutions in place that can handle a wide variety of software; such solutions may be built into operating systems or provided through third-party software installed on their client devices and other WLAN components. Also, WLAN client management software often has configuration management capabilities; this software may already be installed on the WLAN devices. If the existing configuration management solutions cannot collectively provide the necessary WLAN configuration implementation and maintenance capabilities, then the organization should evaluate alternatives and implement a suitable configuration management solution that provides the missing capabilities.

Organizations should evaluate all standardized WLAN security configuration implementations carefully before deploying them throughout production environments.

Even if the organization is confident that the configuration is sound, it should still be evaluated carefully to ensure that its implementation meets the organization's particular security and operational requirements.

Every operating environment has unique characteristics that should be taken into account. It is particularly important to evaluate the strength of the configuration and its potential impacts on performance and functionality (for example, a security setting that is incompatible with use of a particular application).

3. WLAN SECURITY MONITORING

This section discusses two types of security monitoring: assessments [SP800-115, SP800-53A] and continuous monitoring. Security assessment is "the process of determining how effectively an entity being assessed meets specific security objectives."

Security assessments are typically performed periodically, such as annually or quarterly, and are often called periodic assessments. Continuous monitoring is defined in [SP800-137] as "maintaining ongoing awareness of information security, vulnerabilities, and threats to support organizational risk management decisions.

The objective is to conduct ongoing monitoring of the security of an organization's networks, information, and systems, and respond by accepting, avoiding/rejecting, transferring/sharing, or mitigating risk as situations change." Think of continuous monitoring as doing assessments in near real-time, so that the organization can react to problems in minutes instead of months or years.

For WLAN security, the same tools are often useful for both periodic assessments and continuous monitoring. These forms of monitoring are used to ensure that each WLAN component's security configuration meets the organization's requirements (and to report all deviations from these requirements), as well as to monitor WLAN activity to identify attacks, improperly secured communications, and other WLAN security issues. This section provides recommendations for WLAN security monitoring (both periodic assessments and continuous monitoring).

As explained in Section 1.2, the recommendations presented in this section supplement those provided for specific WLAN technologies by other NIST publications [SP800-48, SP800-97], as well as NIST publications on general security assessment [SP800-115, SP800-53A] and continuous monitoring [SP800-137].

3.1. WLAN Security Monitoring Basics

Security monitoring is important for all systems and networks, but it is generally even more important for WLANs. WLANs and wired networks have the same security objectives and face the same general types of threats, but WLAN security is typically harder to achieve for several reasons.

First, WLANs tend to be easier for attackers to access than wired networks. The most significant difference between protecting WLANs and wired networks is the relative ease of intercepting wireless network transmissions and inserting new or altered transmissions from what is presumed to be the authentic source. To monitor traffic on a wired network, an attacker would have to gain physical access to the network or remotely compromise systems on the network; for a WLAN, an attacker simply needs to be within range of the wireless transmissions. (Some attackers use highly sensitive directional antennas, which can greatly extend the effective range of attack beyond the standard WLAN range.)

Second, WLANs are often poorly secured. For example, many WLANs are configured so that they do not require strong authentication; this makes it much easier, sometimes trivial, for attackers within range of the WLAN to successfully gain access to it. These configurations are often used because they are more convenient for users and administrators, but they often put the transmitted information and WLAN devices at serious risk of compromise.

Third, WLANs are usually connected to an organization's wired networks. This means that the WLANs and WLAN devices are not only subject to WLAN-specific attacks, but also nearly all the attacks that wired networks and devices on those networks face.

To support WLAN security, organizations should perform both attack monitoring and vulnerability monitoring. Sections 3.1.1 and 3.1.2 provide more information on these, respectively.

3.1.1. Attack Monitoring

Organizations should continuously monitor their WLANs for both WLAN-specific and general (wired network) attacks. The latter involves the same security controls as would be used for any network-connected system in an organization; see [SP800-53] for additional information on general system security and [SP800-94] for general information on attack detection technologies. Further discussion of general (non-wireless) attack monitoring is out of the scope of this publication.

WLAN-specific attacks can typically be divided into two types: passive and active. These attack classes, which are significant for monitoring purposes, are described below.

- **Passive attack:** an attack in which an unauthorized party only monitors WLAN communications; the attacker does not generate, alter, or disrupt WLAN communications. There are two types of passive attacks:
  - **Eavesdropping.** The attacker monitors WLAN data transmissions for message content.
  - **Traffic analysis** (also known as traffic flow analysis). The attacker gains intelligence by monitoring the transmissions for patterns of communication. A considerable amount of information is contained in the flow of messages between communicating parties.
- **Active attack:** an attack in which an unauthorized party generates, alters, or disrupts WLAN communications. Active attacks may take the form of one of the following types:
  - **Masquerading.** The attacker impersonates an authorized user to gain access to certain unauthorized privileges.
  - **Replay.** The attacker monitors transmissions (passive attack) and retransmits messages posing as the legitimate user.
  - **Message modification.** The attacker alters a legitimate message by deleting, adding to, changing, or reordering the message.
  - **Denial of service (DoS).** A DoS can occur inadvertently, such as other electronic devices causing interference, or it can occur intentionally, such as an attacker sending large numbers of messages at a high rate to flood the WLAN.
  - **Misappropriation.** The attacker steals or makes other unauthorized use of WLAN services.

A form of active attack that is particularly noteworthy is the deployment of rogue WLAN devices. For example, an attacker deploys an AP that has been configured to appear as part of an organization's WLAN infrastructure. This can provide a backdoor into the wired network, bypassing perimeter security mechanisms, such as firewalls. In addition, if client devices inadvertently connect to the rogue AP, the attacker can view and manipulate the client devices' communications (e.g., man-in-the-middle attacks), as well as potentially gaining access to the client devices themselves.

Monitoring WLAN-specific attacks is focused on active attacks. Because passive attacks strictly involve intercepting radio transmissions and do not generate their own transmissions, there is nothing for an organization to monitor electronically. However, all forms of active attack mentioned above can be detected through monitoring. Section 3.2 discusses tools that can monitor for active attacks.

3.1.2. Vulnerability Monitoring

Organizations should do largely the same vulnerability monitoring for WLAN components that they do for any other software: identifying patches and applying them, and verifying security configuration settings and adjusting them as needed (see Section 2). These actions should be performed at least as often for WLAN components as they are for the organization's equivalent wired systems. Further discussion of these general vulnerability monitoring techniques is outside the scope of this publication.

In addition to the general vulnerability monitoring, WLAN-specific vulnerability monitoring involves analyzing WLAN communications and identifying policy violations, such as communications using the wrong protocols, encryption key lengths, etc. This is effective at identifying misconfigured WLAN devices as well as devices that are acting differently than their configuration indicates. It is especially helpful when not all of the WLAN devices are under the organization's control, such as visitor laptops, and when unauthorized WLAN devices are a concern (e.g., an employee attaching a personal mobile device to an organization WLAN without authorization). This form of vulnerability monitoring is often provided by the same tools that are effective at attack monitoring, as explained in Section 3.2.

3.2. Monitoring Tools

One of the primary tools for WLAN security monitoring is a wireless intrusion detection and prevention system (WIDPS). A WIDPS has sensors placed at designated locations within an organization's facilities; these sensors monitor WLAN bands and channels to sample traffic, allowing them to identify WLAN attacks and some WLAN vulnerabilities. WIDPS sensors are available in several forms:

- **Dedicated.** A dedicated sensor performs WIDPS functions but does not pass network traffic from source to destination. Dedicated sensors are often completely passive, simply sniffing WLAN traffic. Some dedicated sensors analyze the traffic they monitor, while other sensors forward the traffic to a management server for analysis. The sensor is typically connected to a wired network. Dedicated sensors are usually designed for one of two deployment types:
  - **Fixed**—the sensor is deployed to a particular location. Fixed sensors usually depend on the organization's infrastructure for power, wired network access, and other resources.
  - **Mobile**—the sensor is designed to be portable so it can be used from multiple locations or while in motion. For example, a security administrator could use a mobile sensor while walking through an organization's buildings to find rogue APs.
- **Bundled.** Many APs and wireless switches offer some WIDPS capabilities as a secondary function.

In addition to these hardware-based forms, there are also host-based WIDPS sensor software products that can be installed on WLAN client devices. The sensor software detects WLAN attacks within range of the client device, as well as WLAN vulnerabilities of the client device, and reports this information to WIDPS management servers. The sensor software may also be able to enforce WLAN security policies. See [SP800-94] for more information on all forms of WIDPS.

Another primary tool for WLAN security monitoring is a WLAN scanner. A passive WLAN scanning tool captures WLAN traffic being transmitted within the range of the tool's antenna. Most passive tools record the key attributes of discovered WLAN devices. This information can be used to identify WLAN vulnerabilities and to detect potential rogue devices. WLAN scanning tools used to conduct completely passive scans transmit no data, nor do the tools in any way affect the operation of deployed WLAN devices. By not transmitting data, a passive scanning tool remains undetected by malicious users and other devices. This reduces the likelihood of individuals avoiding detection by disconnecting or disabling unauthorized WLAN devices.

There are also active WLAN scanning tools. Building on the information collected during passive scans, active scanning attempts to attach to discovered WLAN devices and conduct penetration or vulnerability-related testing. Organizations should be cautious in conducting active scans to make sure they do not inadvertently scan devices owned or operated by other organizations or individuals within range. It is important to evaluate the physical location of devices before actively scanning them.

Generally, organizations should focus on identifying and locating potential rogue devices rather than performing active scans of such devices.

The types of tools mentioned in this section are examples of tools that are often effective at WLAN security monitoring. This is not intended to imply that no other types of tools may also be useful at WLAN security monitoring. What is important is having the necessary monitoring capabilities, not the labels applied to the tool or tools providing those capabilities.

3.3. Continuous Monitoring Recommendations

Organizations with WLANs should implement continuous monitoring solutions for their WLANs that provide all of the following detection capabilities:

- Unauthorized WLAN devices, including rogue APs and unauthorized client devices
- WLAN devices that are misconfigured or using weak WLAN protocols and protocol implementations
- Unusual WLAN usage patterns, such as extremely high numbers of client devices using a particular AP, abnormally high volumes of WLAN traffic involving a particular client device, or many failed attempts to join the WLAN in a short period of time
- The use of active WLAN scanners (e.g., war driving tools) that generate WLAN traffic. The use of passive sensors cannot be detected through monitoring controls.
- DoS attacks and conditions (e.g., network interference). Many denial of service attacks are detected by counting events during periods of time and alerting when threshold values are exceeded. For example, a large number of events involving the termination of WLAN sessions can indicate a DoS attack.
- Impersonation and man-in-the-middle attacks. For example, some WIDPS sensors can detect when a device is attempting to spoof the identity of another device.

Organizations with WLANs should also have the capability to identify the physical location of a detected WLAN threat by using triangulation—

estimating the threat's approximate distance from multiple sensors by the strength of the threat's signal received by each sensor, then calculating the physical location at which the threat would be the estimated distance from each sensor. This allows an organization to send appropriate personnel, such as physical security staff, to the location to address the threat.

3.4. Periodic Assessment Recommendations

Organizations with WLANs should conduct regular periodic technical WLAN security assessments. These assessments should be performed at least annually to evaluate the overall security of the WLAN. In addition, organizations should perform periodic assessments at least quarterly unless continuous monitoring of WLAN security is already collecting all of the information about WLAN attacks and vulnerabilities needed for assessment purposes. For example, an organization that does not have comprehensive WIDPS coverage of its facilities should use mobile WIDPS sensors, WLAN scanners, or other tools with similar capabilities to search for rogue WLANs in areas outside the WIDPS's range.

The following are additional factors that organizations should consider when planning the frequency and breadth of periodic assessments:

- The location of the facility being scanned, because the physical proximity of a building to a public area (e.g., streets and public common areas) or its location in a busy metropolitan area may increase the risk of WLAN threats
- The security level of the data to be transmitted on the WLAN
- How often WLAN client devices connect to and disconnect from the environment, and the typical traffic levels for these devices (e.g., occasional activity or fairly constant activity)—this is because only active WLAN client devices are discoverable during a WLAN scan
- Physical changes to the facilities, such as construction projects that could affect the strength and propagation of WLAN signals

APPENDIX A. SUPPORTING NIST SP 800-53 SECURITY CONTROLS AND PUBLICATIONS

The major controls in the NIST Special Publication 800-53, *Recommended Security Controls for Federal Information Systems and Organizations* control catalog that affect an organization's wireless local area network (WLAN) security are:

***AC-3*, Access Enforcement**
Related controls: AC-2, AC-4, AC-5, AC-6, AC-16, AC-17, AC-18, AC-19, AC-20, AC-21, AC-22, AU-9, CM-5, CM-6, MA-3, MA-4, MA-5, SA-7, SC-13, SI-9

***AC-4*, Information Flow Enforcement**
Related controls: AC-17, AC-19, AC-21, CM-7, SA-8, SC-2, SC-5, SC-7, SC-18

***AC-18*, Wireless Access**
Related controls: AC-3, IA-2, IA-3, IA-8
References: NIST Special Publications 800-48, 800-94, 800-97

***AC-19*, Access Control for Mobile Devices**
Related controls: MP-4, MP-5
References: NIST Special Publications 800-114, 800-124

***AC-20*, Use of External Information Systems**
Related controls: AC-3, AC-17, PL-4
References: FIPS Publication 199

***CA-2*, Security Assessments**
Related controls: CA-6, CA-7, PM-9, SA-11
References: FIPS Publication 199; NIST Special Publications 800-37, 800-53A, 800-115

***CA-7*, Continuous Monitoring**
Related controls: CA-2, CA-5, CA-6, CM-3, CM-4

References: NIST Special Publications 800-37, 800-53A; US-CERT Technical Cyber Security Alerts; DOD Information Assurance Vulnerability Alerts

***CM-6*, Configuration Settings**
Related controls: CM-2, CM-3, SI-4
References: OMB Memoranda 07-11, 07-18, 08-22; NIST Special Publications 800-70, 800-128; Web: nvd.nist.gov; www.nsa.gov)

***AU-2*, Auditable Events**
Related control: AU-3
References: NIST Special Publication 800-92

***IA-2*, Identification and Authentication (Organizational Users)**
Related controls: AC-14, AC-17, AC-18, IA-4, IA-5
References: HSPD 12; OMB Memorandum 04-04; FIPS Publication 201; NIST Special Publications 800-63, 800-73, 800-76, 800-78

***IA-3*, Device Identification and Authentication**
Related controls: AC-17, AC-18

***IA-5*, Authenticator Management**
Related controls: AC-2, IA-2, PL-4, PS-6
References: OMB Memorandum 04-04; FIPS Publication 201; NIST Special Publications 800-73, 800-63, 800-76, 800-78

***IA-8*, Identification and Authentication (Non-Organizational Users)**
Related controls: AC-14, AC-17, AC-18, MA-4
References: OMB Memorandum 04-04; Web: www.cio.gov/eauthentication; NIST Special Publication 800-63

***PE-18*, Location of Information System Components**

***RA-3*, Risk Assessment**
Reference: NIST Special Publication 800-30

***RA-5*, Vulnerability Scanning**
Related controls: CA-2, CM-6, RA-3, SI-2

References: NIST Special Publications 800-40, 800-70, 800-115; Web: cwe.mitre.org; nvd.nist.gov

***SC-7*, Boundary Protection**
Related controls: AC-4, IR-4, SC-5
References: FIPS Publication 199; NIST Special Publications 800-41, 800-77

SC-8, Transmission Integrity
Related controls: AC-17, PE-4
References: FIPS Publications 140-2, 197; NIST Special Publications 800-52, 800-77, 800-81, 800-113; NSTISSI No. 7003

SC-9, Transmission Confidentiality
Related controls: AC-17, PE-4
References: FIPS Publications 140-2, 197; NIST Special Publications 800-52, 800-77, 800-113; CNSS Policy 15; NSTISSI No. 7003

***SI-2*, Flaw Remediation**
Related controls: CA-2, CA-7, CM-3, MA-2, IR-4, RA-5, SA-11, SI-11
References: NIST Special Publication 800-40

***SI-4*, Information System Monitoring**
Related controls: AC-4, AC-8, AC-17, AU-2, AU-6, SI-3, SI-7
References: NIST Special Publications 800-61, 800-83, 800-92, 800-94

Information on these controls and guidelines on possible implementations can be found in the following publications:

- *SP 800-30, Risk Management Guide for Information Technology Systems*
- *Draft SP 800-30 Revision 1, Guide for Conducting Risk Assessments*
- *SP 800-37, Guide for Applying the Risk Management Framework to Federal Information Systems: A Security Life Cycle Approach*
- *SP 800-40 Version 2.0, Creating a Patch and Vulnerability Management Program*
- *SP 800-41 Rev. 1, Guidelines on Firewalls and Firewall Policy*
- *SP 800-48 Rev. 1, Guide to Securing Legacy IEEE 802.11 Wireless Networks*

- *SP 800-52, Guidelines for the Selection and Use of Transport Layer Security (TLS) Implementations*
- *SP 800-53 Rev. 3, Recommended Security Controls for Federal Information Systems and Organizations*
- *SP 800-53A Rev. 1, Guide for Assessing the Security Controls in Federal Information Systems and Organizations*
- *SP 800-61 Rev. 1, Computer Security Incident Handling Guide*
- *SP 800-63 Rev. 1: E-Authentication Guideline*
- *SP 800-70 Rev. 2, National Checklist Program for IT Products--Guidelines for Checklist Users and Developers*
- *SP 800-73-3, Interfaces for Personal Identity Verification*
- *Draft SP 800-76-2, Biometric Data Specification for Personal Identity Verification*
- *SP 800-77, Guide to IPsec VPNs*
- *SP 800-78-3, Cryptographic Algorithms and Key Sizes for Personal Identification Verification (PIV)*
- *SP 800-81 Rev. 1, Secure Domain Name System (DNS) Deployment Guide*
- *SP 800-83, Guide to Malware Incident Prevention and Handling*
- *SP 800-92, Guide to Computer Security Log Management*
- *SP 800-94, Guide to Intrusion Detection and Prevention Systems (IDPS)*
- *SP 800-97, Establishing Wireless Robust Security Networks: A Guide to IEEE 802.11i*
- *SP 800-113, Guide to SSL VPNs*
- *SP 800-114, User's Guide to Securing External Devices for Telework and Remote Access*
- *SP 800-115, Technical Guide to Information Security Testing and Assessment*
- *SP 800-124, Guidelines on Cell Phone and PDA Security*
- *SP 800-128, Guide for Security-Focused Configuration Management of Information Systems*

APPENDIX B. ACRONYMS AND ABBREVIATIONS

Selected acronyms and abbreviations used in the guide are defined below.

AP	Access Point
BIOS	Basic Input/Output System
DHS	Department of Homeland Security
DISA	Defense Information Systems Agency
DoS	Denial of Service
DS	Distribution System
FISMA	Federal Information Security Management Act
GAO	Government Accountability Office
IDPS	Intrusion Detection and Prevention System
IEEE	Institute of Electrical and Electronics Engineers
IT	Information Technology
ITL	Information Technology Laboratory
LAN	Local Area Network
NIST	National Institute of Standards and Technology
OMB	Office of Management and Budget
OS	Operating System
SP	Special Publication
STIG	Security Technical Implementation Guide
TIC	Trusted Internet Connection
USB	Universal Serial Bus
WIDPS	Wireless Intrusion Detection and Prevention System
WiMAX	Worldwide Interoperability for Microwave Access
WLAN	Wireless Local Area Network

APPENDIX C. REFERENCES

The list below provides references for the publication.

[1] [GAO-11-43] GAO-11-43, *Federal Agencies Have Taken Steps to Secure Wireless Networks, but Further Actions Can Mitigate Risk*, November 2010. http://www.gao.gov/new.items/d1143.pdf

[2] [NCP] National Checklist Program Repository. http://checklists.nist.gov/

[3] [SP800-37] NIST Special Publication 800-37 Revision 1, *Guide for Applying the Risk Management Framework to Federal Information Systems: A Security Life Cycle Approach*, February 2010. http://csrc.nist. gov/publications/PubsSPs.html#800-37

[4] [SP800-48] NIST Special Publication 800-48 Revision 1, *Guide to Securing Legacy IEEE 802.11 Wireless Networks*, July 2008. http://csrc. nist.gov/publications/PubsSPs.html#800-48

[5] [SP800-53] NIST Special Publication 800-53 Revision 3, *Recommended Security Controls for Federal Information Systems and Organizations*, August 2009. http://csrc.nist.gov/publications/PubsSPs.html#800-53

[6] [SP800-53A] NIST Special Publication 800-53A, *Guide for Assessing the Security Controls in Federal Information Systems and Organizations: Building Effective Security Assessment Plans*, June 2010. http://csrc.nist.gov/publications/PubsSPs.html#800-53a

[7] [SP800-70] NIST Special Publication 800-70 Revision 2, *National Checklist Program for IT Products—Guidelines for Checklist Users and Developers*, February 2011. http://csrc.nist.gov/publications/PubsSPs. html#800-70

[8] [SP800-94] NIST Special Publication 800-94, *Guide to Intrusion Detection and Prevention Systems (IDPS)*, February 2007. http://csrc. nist.gov/publications/PubsSPs.html#800-94

[9] [SP800-97] NIST Special Publication 800-97, *Establishing Robust Security Networks: A Guide to IEEE 802.11i*, February 2007. http://csrc. nist.gov/publications/PubsSPs.html#800-97

[10] [SP800-115] NIST Special Publication 800-115, *Technical Guide to Information Security Testing and Assessment*, September 2008. http:// csrc.nist.gov/publications/PubsSPs.html#800-115

[11] [SP800-137] NIST Special Publication 800-137, *Information Security Continuous Monitoring for Federal Information Systems and Organizations*, September 2011. http://csrc.nist.gov/publications/ PubsSPs.html#800-137

End Note

[1] See [GAO-11-43] for additional information on the history of the IEEE 802.11 standard for WLANs.

INDEX

S

T

U

V

W

Y